1 MONTH OF
FREE
READING

at

www.ForgottenBooks.com

By purchasing this book you are eligible for one month membership to ForgottenBooks.com, giving you unlimited access to our entire collection of over 1,000,000 titles via our web site and mobile apps.

To claim your free month visit: www.forgottenbooks.com/free1249679

ISBN 978-0-428-62046-2
PIBN 11249679

for information systems –

computer graphics –
metafile for the
storage and transfer of
picture description information

american national standards institute, inc
1430 broadway, new york, new york 10018

American National Standard

Approval of an American National Standard requires verification by ANSI that the requirements for due process, consensus, and other criteria for approval have been met by the standards developer.

Consensus is established when, in the judgment of the ANSI Board of Standards Review, substantial agreement has been reached by directly and materially affected interests. Substantial agreement means much more than a simple majority, but not necessarily unanimity. Consensus requires that all views and objections be considered, and that a concerted effort be made toward their resolution.

The use of American National Standards is completely voluntary; their existence does not in any respect preclude anyone, whether he has approved the standards or not, from manufacturing, marketing, purchasing, or using products, processes, or procedures not conforming to the standards.

The American National Standards Institute does not develop standards and will in no circumstances give an interpretation of any American National Standard. Moreover, no person shall have the right or authority to issue an interpretation of an American National Standard in the name of the American National Standards Institute. Requests for interpretations should be addressed to the secretariat or sponsor whose name appears on the title page of this standard.

CAUTION NOTICE: This American National Standard may be revised or withdrawn at any time. The procedures of the American National Standards Institute require that action be taken to reaffirm, revise, or withdraw this standard no later than five years from the date of approval. Purchasers of American National Standards may receive current information on all standards by calling or writing the American National Standards Institute.

This standard has been adopted for Federal Government use.

Details concerning its use within the Federal Government are contained in Federal Information Processing Standards Publication 128, Computer Graphics Metafile (CGM). For a complete list of the publications available in the Federal Information Processing Standards Series, write to the Standards Processing Coordinator (ADP), Institute for Computer Sciences and Technology, National Bureau of Standards, Gaithersburg, MD 20899.

Published by

American National Standards Institute
1430 Broadway, New York, New York 10018

Printed in the United States of America

PC2M687/30

American National Standard
for Information Systems –

Computer Graphics –
Metafile for the
Storage and Transfer of
Picture Description Information

Secretariat

Computer and Business Equipment Manufacturers Association

Approved August 27, 1986

American National Standards Institute, Inc

Abstract

The Computer Graphics Metafile (CGM) is a set of basic elements for a computer graphics data interface usable by many graphics-producing systems and applications. This standard (1) allows graphics data to be easily transported between computer graphics devices and installations, (2) aids computer graphics software implementors in understanding and using graphics data storage methods, and (3) guides device manufacturers on useful graphics capabilities.

This standard also defines a presentation-level interface to a graphics system. Hence, it contains elements for (1) graphical output primitives, (2) control of the appearance of graphical primitives with attributes, (3) definition of information important to metafile interpreters, and (4) specification of the parameter modes of attribute elements.

(This Foreword is not part of American National Standard X3.122-1986.)

This American National Standard provides a set of basic elements for computer graphics data. These functions taken as a whole are called the Computer Graphics Metafile (CGM). The design of this standard is based on the work of many groups. Much of the early design methodology was heavily influenced by the work of the Graphics Standards Planning Committee of the Special Interest Group on Computer Graphics of the Association for Computing Machinery (ACM-SIGGRAPH GPSC). This work, known as the Core System, contained a metafile proposal and was widely distributed in 1979. The CGM itself was originally developed by Technical Committee X3H3 in 1981 and subsequently was refined extensively from 1982 to 1984 in cooperation with Working Group 2 of the Subcommittee on Programming Languages of the Technical Committee on Information Processing of the International Standards Organization (ISO TC97/SC21/WG2).

This standard was developed by Technical Committee X3H3 (Computer Graphics Programming Languages) of Accredited Standards Committee X3 operating under the rules and procedures of the American National Standards Institute, under project 347I, started in March 1981. The name of this standard was changed from Virtual Device Metafile (VDM) to Computer Graphics Metafile as a result of a resolution passed by ISO TC97/SC21/WG2 in June 1984. The CGM was approved as an ISO work item (97.5.19) in May 1983. It is currently registered as a DIS and is undergoing DIS balloting within ISO. If any substantive technical corrections occur as a result of the ISO processing, they may be incorporated into this American National Standard at a later date by addenda, corrigenda, or some other means.

This document is limited to the definition of the Computer Graphics Metafile. The CGM is intended to be a graphics data interface. The CGM is closely related in functional capability to the proposed Computer Graphics Interface standard.

The project proposal for the Computer Graphics Metafile (originally the VDM) defined the CGM and its field of application. The following statements from the proposal are repeated here:

   (1) "The VDM is a mechanism for retaining and/or transporting graphics data and control information."

      Retention and transportation may imply the separation, in time or space, or both, of the process that creates the CGM and the process using it. This separation of processes and its implications for assumptions by one process about the characteristics of the other have been considered carefully while producing this metafile standard.

   (2) "This information contains a device-independent description of a picture . . ."

      Device independence inherently limits the set of elements that can be included in a metafile. A standard mechanism (ESCAPE) is provided to permit access to device-dependent features.

   (3) "The VDM is at the level of the Virtual Device Interface."

      The CGM elements are defined consistently with those of the Computer Graphics Interface (CGI) project. (The CGI project was formerly known as the VDI project, its change of name also occurring as a result of the resolution by ISO TC97/SC21/WG2 in June 1984.) This effort defines a minimal, useful set of elements sufficient to describe device-independent graphics pictures. There are several reasons supporting development of this metafile with minimal, but sufficient, capability.

      (a) It encourages early acceptance and implementation

      (b) It is easier to implement

      (c) Its concepts are compatible with existing practice

      (d) It provides a "low-overhead mode of operation"

      Advanced concepts such as segmentation, macro facilities, three-dimensional elements, and others are not included in this standard but are also not precluded from future extensions to this standard.

This standard allows portability of graphics data among installations and encourages a uniform interface for noninteractive picture description.

This standard promotes the exchange of information that enables installations to share work and reduce time spent recomputing in an effort to regenerate graphics data, and enables transfer of data by either media (for example, magnetic tape) or communication (for example, line protocol).

The set of elements comprising the Computer Graphics Metafile uses standardized terminology, thereby allowing both the academic and industrial communities to develop instructional programs that concentrate on programming techniques and methodologies based on these elements.

The major economic benefit is derived from defining a unified external format for graphics data, thus allowing the same graphics data to be displayed on different devices. The following benefits will be derived from this standard:

    (1) Benchmarks can be run on different vendors' equipment

    (2) Graphics output can be recorded as an aid in debugging

    (3) Archiving, off-line plotting, and off-site plotting can take place

    (4) Animation sequences can be built in nonreal time and nonsequential order, and then viewed in real time in the proper sequence

    (5) Selected pictures can be previewed before a large number of pictures are sent to a more expensive or slower medium

    (6) A standard interface can be developed for a variety of plotting, COM, and other off-line, picture-generating devices

    (7) The same picture or series of pictures can be used several times without recalculating the picture

This American National Standard is a graphical picture file exchange standard and not a product definition database exchange standard. Standards work in the latter area is the responsibility of the Y14.26 (Computer-Aided Preparation of Product Definition Data) Technical Committee. This standard is concerned with the generation and transfer of sufficient device-independent information for a picture to be drawn on a wide variety of graphics output devices. The transfer of all product definition data (geometric and non-geometric) across CAD/CAM systems is described in the Initial Graphic Exchange Specification (IGES), which is currently under development as an American National Standard. Specifically, IGES and this standard deal with different information for different purposes at different levels of detail.

Although there are similarities between CGM and the North American Presentation-Level Protocol Syntax (NAPLPS) (see American National Standard Videotex/Teletext Presentation Level Protocol Syntax North American PLPS, ANSI X3.110-1983), the latter is designed to support a particular class of devices in a picture transmission environment, while the CGM is intended to provide picture definition in a device-independent and environment-independent manner.

This American National Standard on CGM has been developed in conjunction with the higher level standards being developed by Technical Committee X3H3. Coordination within Technical Committee X3H3 has taken place in order to achieve consistency among the interrelated standards. The development of the Graphical Kernel System (American National Standard for Information Systems – Computer Graphics – Graphical Kernel System (GKS) Functional Description, ANSI X3.124-1985) and the Computer Graphics Interface (CGI) by Technical Committee X3H3 has been closely related to the development of this American National Standard on metafiles. Also, coordination with work of the X3L2 (Character Sets and Coding), X3J6 (Text Processing Language), and X3J7 (APT) Technical Committees has taken place.

This standard has been developed in collaboration with ISO TC97/SC21/WG2 under project 97.21.5 authorized in May 1983. The International Standard for Information Processing Systems – Computer Graphics – Graphical Kernel System (GKS) Functional Description, ISO 7942-1985, specifically excluded portions pertaining to metafiles in anticipation of this metafile standard. ISO TC97/SC18 (Text Preparation and Interchange) is developing a standard on text imaging capabilities that includes the specification of graphical elements and attributes. Its work has been considered where applicable.

Coordination with the work of ISO TC97/SC2/WG8 (Picture Coding) has taken place. It is expected that a principal use of the CGM will be as a picture capture metafile at level 0a of GKS.

The CGM standard is also being processed through ISO. When approved it will be known as ISO 8632. Throughout ISO 8632 a number of ISO standards are referred to. The following list shows the equivalent American National Standards.

    ISO 646-1983 – ANSI X3.4-1986
    ISO 2022-1982 – ANSI X3.41-1974
    ISO 7942-1985 – ANSI X3.124-1985
    ISO DIS 8632-1987 – ANSI X3.122-1986
    ISO 6429-1983 – ANSI X3.64-1979

American National Standard for Binary Floating-Point Arithmetic (ANSI/IEEE 754-1985) is used for the floating-point representation within the binary encoding.

This standard was approved as an American National Standard by the American National Standards Institute on.August 27, 1986.

Suggestions for improvement of this standard will be welcome. They should be sent to the Computer and Business Equipment Manufacturers Association, 311 First Street, NW, Washington, DC 20001.

This standard was processed and approved for submittal to ANSI by the Accredited Standards Committee on Information Processing Systems, X3. Committee approval of the standard does not necessarily imply that all committee members voted for its approval. At the time it approved this standard, the X3 committee had the following members:

Edward Lohse, Chair
Richard Gibson, Vice-Chair
Catherine A. Kachurik, Administrative Secretary

Organization Represented	Name of Representative
American Express.	D. L. Seigal
	Lucille Durfee (Alt)
American Library Association.	Paul Peters
American Nuclear Society	Geraldine C. Main
	D. R. Vondy (Alt)
AMP Incorporated	Patrick E. Lannan
	Edward Kelly (Alt)
Association for Computing Machinery	Kenneth Magel
	Jon A. Meads (Alt)
Association of the Institute for Certification	
of Computer Professionals.	Thomas M. Kurihara
AT&T Communications	Henry L. Marchese
	Richard Gibson (Alt)
AT&T Technologies	Herbert V. Bertine
	Paul D. Bartoli (Alt)
	Stuart M. Garland (Alt)
Burroughs Corporation.	Stanley Fenner
Control Data Corporation	Charles E. Cooper
	Keith Lucke (Alt)
Cooperating Users of Burroughs Equipment.	Thomas Easterday
	Donald Miller (Alt)

Organization Represented	Name of Representative
Data General Corporation	John Pilat
	Lyman Chapin (Alt)
Data Processing Management Association	Christian G. Meyer
	Ward Arrington (Alt)
	Terrance H. Felker (Alt)
Digital Equipment Computer Users Society	William Hancock
	Dennis Perry (Alt)
Digital Equipment Corporation	Gary S. Robinson
	Delbert L. Shoemaker (Alt)
Eastman Kodak	Gary Haines
	Carleton C. Bard (Alt)
General Electric Company	Richard W. Signor
	William R. Kruesi (Alt)
General Services Administration	William C. Rinehuls
	Larry L. Jackson (Alt)
GUIDE International	Frank Kirshenbaum
	Sandra Swartz Abraham (Alt)
Harris Corporation	Walter G. Fredrickson
	Rajiv Sinha (Alt)
Hewlett-Packard	Donald C. Loughry
Honeywell Information Systems	Thomas J. McNamara
	David M. Taylor (Alt)
IBM Corporation	Mary Anne Gray
	Robert H. Follett (Alt)
IEEE Computer Society	Sava I. Sherr
	Thomas M. Kurihara (Alt)
	Thomas A. Varetoni (Alt)
Lawrence Berkeley Laboratory	David F. Stevens
	Robert L. Fink (Alt)
Moore Business Forms	Delmer H. Oddy
National Bureau of Standards	Robert E. Rountree
	James H. Burrows (Alt)
National Communications System	George W. White
NCR Corporation	Thomas W. Kern
	A. Raymond Daniels (Alt)
Prime Computer, Inc	Joseph Schmidt
	John McHugh (Alt)
Railinc Corporation	R. A. Petrash
Recognition Technology Users Association	Herbert F. Schantz
	G. W. Wetzel (Alt)
SHARE, Inc	Thomas B. Steel
	Robert A. Rannie (Alt)
Sperry Corporation	Marvin W. Bass
	Jean G. Smith (Alt)
Texas Instruments, Inc	Presley Smith
	Richard F. Trow, Jr (Alt)
3M Company	Paul D. Jahnke
	J. Wade Van Valkenburg (Alt)
Travelers Insurance Companies, Inc.	Joseph T. Brophy
U.S. Department of Defense	Fred Virtue
	Belkis Leong-Hong (Alt)
VIM	Chris Tanner
	Madeline Sparks (Alt)
VISA U.S.A	Jean T. McKenna
	Susan Crawford (Alt)
Wang Laboratories, Inc.	Marsha Hayek
	Joseph St. Amand (Alt)
Xerox Corporation	John L. Wheeler
	Roy Pierce (Alt)

Technical Committee X3H3 on Computer Graphics, which developed the draft proposals, which held the U.S. Technical Advisory Group responsibilities for ISO TC97/SC21/WG2, and through which this standard was completed, had the following members at the time of the first Public Review of this standard:

P. Bono, Chair
B. Shepherd, Vice-Chair
R. Simons, Secretary
J. Chin, International Representative

D. Bailey
K. Hepworth (Alt)
J. Bedrick
W. Yip (Alt)
B. Sangster (Alt)
J. Blair
P. Bono
W. Brown
A. Bunshaft
J. Butler
D. McCabe (Alt)
W. Dale
R. Ehlers
F. Canfield
S. Carson
J. Chin
G. Cuthbert
S. Gill
D. Slaby (Alt)
J. Hargrove
D. Galewsky (Alt)
K. Kimbrough (Alt)
R. Harney
M. Heck
R. Bruns (Alt)
M. Languth (Alt)
M. Plaehn (Alt)
L. Henderson
W. Johnston
D. Cahn (Alt)
P. Jones
M. Babcock (Alt)
A. Leinwand (Alt)
M. Journey
S. Stash (Alt)
J. Kearney
R. Flippen (Alt)
F. Langhorst
R. Kan (Alt)
O. Lapczak
D. Lynch
K. Leung (Alt)
J. Reese (Alt)
C. Mannhardt
T. Morrissey

A. Frankel (Alt)
E. McGinnis (Alt)
P. Showman (Alt)
H. Newman
P. Norman
B. Perry
T. Clarkson (Alt)
E. Post (Alt)
B. Plunkett
T. Powers
J. McConnell (Alt)
R. McNall (Alt)
L. Preheim
R. Holzman (Alt)
R. Puk
T. Reed
R. Elliott (Alt)
J. Rowe
K. Schmucker
C. Seum
B. Cohen (Alt)
C. Duffy (Alt)
B. Olenchuk (Alt)
J. Schoenburg (Alt)
M. Skall
B. Shepherd
A. Herrick (Alt)
R. Simons
D. Shuey
E. Sonderegger
L. Hatfield (Alt)
N. Soong
W. Hafner (Alt)
T. Mainock (Alt)
M. Sparks
R. Stout
L. Benbrooks (Alt)
T. Thornton (Alt)
D. Straayer
B. Ross (Alt)
G. Strockbine
M. Whyles
T. Wright
D. Kusumoto (Alt)

Document support for this standard was provided by National Center for Atmospheric Research, Boulder, Colorado.

nts Part 1 Functional Specification

Part 2 Character Encoding

Part 3 Binary Encoding

Part 4 Clear Text Encoding

ANSI X3.122 - 1986

Information Processing Systems

Computer Graphics

Metafile for the Storage and Transfer
of Picture Description Information

Part 1

Functional Specification

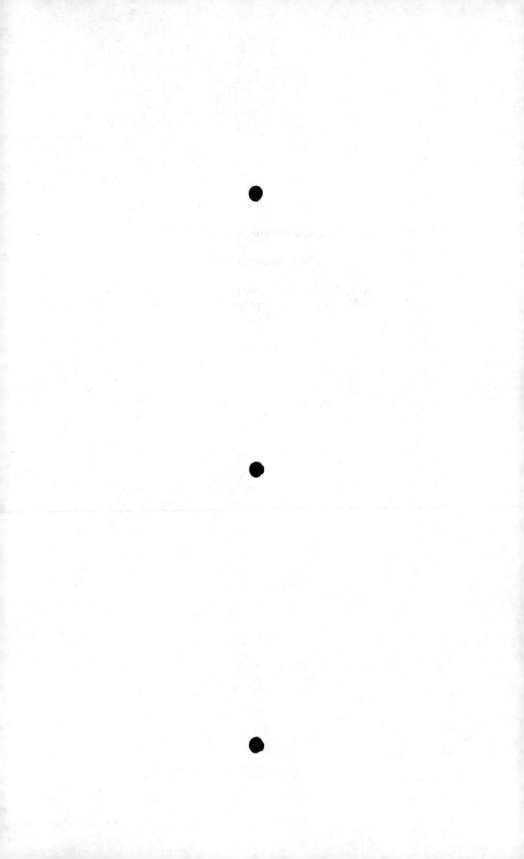

ANSI X3.122 - 1986
Part 1

CONTENTS

American National Standard
for Information Systems –

Computer Graphics –
Metafile for the
Storage and Transfer of
Picture Description Information

0 Introduction

0.1 Purpose

The Computer Graphics Metafile provides a file format suitable for the storage and retrieval of picture information. The file format consists of a set of elements that can be used to describe pictures in a way that is compatible between systems of different architectures and devices of differing capabilities and design.

0.2 Reasons for this Standard

The main reasons for producing a standard computer graphics metafile are:

a) to allow picture information to be stored in an organized way on a graphical software system;

b) to facilitate transfer of picture information between different graphical software systems;

c) to enable picture information to be transferred between graphical devices;

d) to enable picture information to be transferred between different computer graphics installations.

0.3 Design Requirements

To reach these objectives, a number of design principles were adopted:

a) The Metafile should provide a suitable set of elements for the transfer of a wide range of pictorial information.

b) The Metafile should address the more usual and essential features found on graphical devices directly and should provide access to less common facilities via an escape mechanism.

c) The design of the Metafile should not preclude extension of the Standard at a later stage to cover facilities beyond those included in this version of the Standard.

d) The Metafile should be usable from GKS (Graphical Kernel System - ISO 7942) with both Metafile Input and Metafile Output functions.

e) The Standard should address the needs of different applications that have conflicting requirements for size of metafile, speed of generation and interpretation, readability, editability and ease of transfer through different transport mechanisms.

0.4 Design Criteria

The requirements of sub-clause 0.3 were used to formulate the following criteria which were used to decide between different design possibilities.

ANSI X3.122 - 1986 Part 1

a) Completeness: In any area of the Standard, the functionality specified by the Standard should be complete in itself.

b) Conciseness: Redundant elements or parameters should be avoided.

c) Consistency: Contradictory elements should be avoided.

d) Extensibility: The ability to add new elements and generality to the Standard should not be precluded.

e) Fidelity: The minimal results and characteristics of elements should be well defined.

f) Implementability: An element should be able to be supported efficiently on most host systems and/or graphics hardware.

g) Orthogonality: The elements of the metafile should be independent of each other, or any dependencies should be structured and well defined.

h) Predictability: The Standard should be such that the recommended or proper use of standard elements guarantees the results of using a particular element.

i) Standard practice: Only those elements that reflect existing practice, that are necessary to support existing practice, or that are necessary to support proposed standards should be standardized.

j) Usefulness: Functions should be powerful enough to perform useful tasks.

k) Well-structured: The assumptions that elements make about each other should be minimized. An element should have a well-defined interface and a simply stated unconditional purpose. Multipurpose elements and side effects should be avoided.

0.5 Access to a Metafile

The Metafile has been designed so that, although its main usage is anticipated as being with completely sequential access, non-sequential access is also possible. Once the basic environment of the metafile has been established, individual pictures may be accessible if the medium, the encoding and the implementation support this form of access.

0.6 Generation and Interpretation of Metafiles

The specific mechanisms of metafile generation and interpretation are not described by this Standard, although it does describe the intended result of such interpretation. The basic set of Metafile elements includes a capability for the addition of application-dependent data, which do not have graphical meaning and for which no intended interpretation results are described.

0.7 Distinction between Formal Specification and Encodings

The functionality provided by the Metafile is separated from the specification of any particular encoding format. This Standard provides for both standard and private encodings of the elements described in part 1, "Functional Specification." Guidelines for private encodings are specified in part 1, annex b; these guidelines are not part of this Standard.

Three standard encodings are specified in parts 2, 3 and 4. Each of the standardized encodings is capable of representing the full functionality described in part 1 of this Standard. Translation between the standardized encodings is possible without loss of picture information, although subsequent translation back into the original encoding may not result in precisely the same data stream, due to different quantizations of precisions in the different encodings.

The Character Encoding specified in part 2 is intended to provide an encoding of minimum size. It conforms to the rules for code extension specified in ISO 2022 in the category of complete code system. It is particularly suitable for transfer through networks that cannot support binary transfers.

The Binary Encoding specified in part 3 provides an encoding that requires least effort to generate and interpret on many systems.

The Clear Text Encoding specified in part 4 provides an encoding that can be created, viewed and edited with standard text editors. It is therefore also suitable for transfer through networks that support only transfer of text files.

0.8 Relationship to Other ISO Standards

This Standard draws extensively for its model of a graphics system on GKS (Graphical Kernel System — ISO 7942). In addition, this Standard specifies a metafile that may be used as a static picture-capture metafile by GKS. One relationship between this Standard and GKS — the use of a subset of the elements of this Standard as a static picture capture metafile by GKS — is explained in annex E.

The Character Encoding specified in part 2 conforms to the code extension techniques of ISO 2022.

The Binary Encoding specified in part 3 employs the mechanism for representing floating point numbers specified in ANSI/IEEE 754-1986.

For certain elements, the CGM defines value ranges of parameters as being reserved for registration. The meanings of these values will be defined using the established procedures (see 4.11.) the ISO International Registration Authority for Graphical Items. These procedures do not apply to values and value ranges defined as being reserved for private use; these values and ranges are not standardized.

0.9 Status of Annexes

In all parts of this Standard, the annexes do not form an integral part of the Standard but are included to provide extra information and explanation.

1 Scope and Field of Application

The Computer Graphics Metafile provides a file format suitable for the storage and retrieval of picture description information. The file format consists of an ordered set of elements that can be used to describe pictures in a way that is compatible between systems of different architectures and devices of differing capabilities and design.

The elements specified provide for the representation of a wide range of pictures on a wide range of graphical devices. The elements are split into groups that delimit major structures (metafiles and pictures), that specify the representations used within the metafile, that control the display of the picture, that perform basic drawing actions, that control the attributes of the basic drawing actions and that provide access to non-standard device capabilities.

The Metafile is defined in such a way that, in addition to sequential access to the whole metafile, random access to individual pictures is well-defined; whether this is available in any system that uses this Standard depends on the medium, the encoding and the implementation.

In addition to a Functional Specification, three standard encodings of the metafile syntax are specified. These encodings address the needs of applications that require minimum metafile size, minimum effort to generate and interpret, and maximum flexibility for a human reader or editor of the metafile.

Part 1 of this Standard describes the format using an abstract syntax. The remaining three parts specify three standardized encodings that conform to this syntax: part 2 specifies a character encoding that conforms to the rules for code extension specified in ISO 2022 — Code Extension Techniques — in the category of complete coding system; part 3 specifies a binary encoding; part 4 specifies a clear text encoding.

2 References

ISO 646	Information Processing - 7-bit coded character set for information interchange
ISO 2022	Information Processing - ISO 7-bit and 8-bit coded character sets - Code Extension techniques
ISO 2375	Information Processing - Character Set Registration
ISO 7942	Information Processing Systems - Computer Graphics - Functional Specification of the Graphical Kernel System (GKS)

3 Definitions and Abbreviations

aspect ratio: The ratio of the width to the height of a rectangular area, such as a window or viewport. For example, an aspect ratio of 2.0 indicates an area twice as wide as it is high.

aspect source flag (ASF): Indicator as to whether a particular attribute selection is to be individual or bundled.

aspects of primitives: Ways in which the appearance of a primitive can vary. Some aspects are controlled directly by primitive attributes; some can be controlled indirectly through a bundle table.

attribute elements: Metafile elements that describe the appearance of graphical elements.

bundle: Set of attributes associated with one of the following graphical element types: line, marker, text, and filled area.

bundle index: Index for accessing a particular set of attributes in a bundle table.

bundle table: An indexed table containing a set of attributes for each index.

clip indicator: Indicator as to whether metafile graphical elements are to be clipped at the limits of CLIP RECTANGLE.

clip rectangle: A rectangle defined in VDC space which is used as a clipping boundary when the metafile graphical elements are to be clipped.

clipping: The process of removing any portion of a graphical image which extends beyond a specified boundary.

colour selection mode: Indicator as to whether colour selection is to be direct (by specifying the RGB values) or indexed (by specifying an index into a table of RGB values).

colour table: A table for use in mapping from a colour index to the corresponding colour. See DIRECT COLOUR, INDEXED COLOUR.

colour value: The values of the RGB (red, green, blue) components describing a colour.

Computer Graphics Interface (CGI): The specification for interface techniques for dialogues with graphical devices.

Computer Graphics Metafile (CGM): The specification for a mechanism for storing and transferring picture description information.

conjugate diameter pair (CDP): A pair D,d of diameters of an ellipse such that a tangent to the ellipse at each endpoint of one diameter is parallel to the other diameter.

control elements: Metafile elements that specify metafile delimiters, address space, clipping boundaries, picture delimiters, and format descriptions of the Metafile elements.

data interface: An interface between software modules or devices comprising one or more packets containing opcodes and data—as contrasted with a subroutine call interface.

descriptor elements: Metafile elements that describe the functional content, format, default conditions, identification, and characteristics of the Metafile.

device driver: The device-dependent part of a graphics implementation which supports a physical device. The device driver generates device-dependent output.

direct colour: A colour selection scheme in which the colour values are specified directly, without requiring an intermediate mapping via a colour table. See COLOUR TABLE, INDEXED COLOUR.

display surface: That part of a graphics device upon which a visible image appears (for example, the screen of a display, the paper in a plotter).

escape elements: Metafile elements that describe device- or system-dependent elements used to construct a picture, but that are not otherwise standardized.

external elements: Metafile elements that communicate information not directly related to the generation of a graphical image.

font: As used in this Standard, the typeface or style of characters, independent of other text attributes such as size and rotation. The font is distinct from the character set.

graphical elements: Metafile elements that describe images in the Metafile.

Graphical Kernel System (GKS): A standardized application programmer's interface to graphics.

graphics device: A device (for example, refresh display, storage tube display, or plotter) on which display images can be represented.

hatch style: A format for filling closed figures. A hatch style consists of one or more sets of lines whose presence represents the interior of the figure in question.

indexed colour: A colour selection scheme in which the colour index is used to retrieve colour values from a colour table. See COLOUR TABLE, DIRECT COLOUR.

message: A string of characters used to communicate information to operators at Metafile interpretation time.

metafile: A mechanism for retaining and transporting graphical data and control information. This information contains a device-independent description of one or more pictures.

Metafile Descriptor (MD): A metafile element that describes the format of the metafile (but not its encoding method) and the functionality expected of a metafile interpreter.

metafile element: A functional item that can be used to construct a picture or convey information.

metafile generator: The process or equipment that produces the Computer Graphics Metafile.

metafile interpreter: The process or equipment that reads the Computer Graphics Metafile and interprets the contents. An interpreter may be needed in order to drive a Computer Graphics Interface or other device interface to obtain a picture that resembles the intended picture as closely as possible.

normalized device coordinates (NDC): Coordinates specified in a device-independent coordinate system, normalized to some range (typically 0 to 1). See VDC EXTENT, VDC RANGE, VDC SPACE, VIRTUAL DEVICE COORDINATES.

pattern style: A format for filling closed figures with patterns. A pattern style consists of an array of variously coloured or shaded cells.

Picture Descriptor (PD): A set of metafile elements used to set the interpretation modes of attribute elements for the entire picture.

pixel: The smallest element of a display surface that can be independently assigned colour.

realized edge: the zero-width ideal boundary line of the filled-area if the edge is invisible, and the finite-width displayed line if the edge is visible.

realized interior: in a filled area element, that portion of the ideal interior as extending to and terminating at the realized edge.

view surface: See DISPLAY SURFACE.

virtual device: An idealized graphics device that presents a set of graphics capabilities to graphics software or systems via the Computer Graphics Interface.

virtual device coordinates (VDC): The coordinates used to specify position in the VDC space. These are absolute two-dimensional coordinates. See VDC SPACE.

VDC extent: A rectangular region of interest contained within the VDC range. See VDC RANGE, VDC SPACE.

VDC range: A rectangular region within VDC space consisting of the set of all coordinates representable in the declared coordinate type, precision, and encoding format of the metafile. See VDC EXTENT, VDC SPACE.

VDC space: A two-dimensional Cartesian coordinate space of infinite precision and extent. Only a subset of VDC space, the VDC range, is realizable in a metafile. See VDC EXTENT, VDC RANGE, VIRTUAL DEVICE COORDINATES.

The following abbreviations are used in all parts of this Standard.

ASF	Aspect Source Flag
CDP	Conjugate Diameter Pair
CGI	Computer Graphics Interface
CGM	Computer Graphics Metafile
GKS	Graphical Kernel System
MD	Metafile Descriptor
NDC	Normalized Device Coordinate(s)
PD	Picture Descriptor
VDC	Virtual Device Coordinate(s)

4 Concepts

4.1 Introduction

The objective of the Computer Graphics Metafile (CGM) is to provide for the description, storage, and communication of graphical information in a device-independent manner. To accomplish this, the Standard defines the form (syntax) and functional behaviour (semantics) of a set of elements that may occur in the CGM. The following classes of elements are defined:

— Delimiter Elements, which delimit significant structures within the Metafile.

— Metafile Descriptor Elements, which describe the functional content, default conditions, identification, and characteristics of the CGM.

— Picture Descriptor Elements, which set the interpretation modes of attribute elements for each picture.

— Control Elements, which allow picture boundaries and coordinate representation to be modified.

— Graphical Primitive Elements, which describe the visual components of a picture in the CGM.

— Attribute Elements, which describe the appearance of graphical primitive elements.

— Escape Element, which describes device- or system-dependent elements used to construct a picture; however, the elements are not otherwise standardized.

— External Elements, which communicate information not directly related to the generation of a graphical image.

A Computer Graphics Metafile is a collection of elements from this standardized set. The BEGIN METAFILE and END METAFILE elements each occur exactly once in a complete metafile; as many or as few of the elements in the other classes may occur as are needed. A metafile needs to be interpreted in order to display its pictorial content on a graphics device. The Descriptor Elements give the interpreter sufficient data to interpret metafile elements and to make informed decisions concerning the resources needed for display.

Any CGM contains certain delimiter elements; in addition it may include control elements for metafile interpretation, Picture Descriptor elements for declaring parameter modes of attribute elements, graphical primitive elements for defining graphical entities, attribute elements for defining the appearance of the graphical primitive elements, escape elements for accessing non-standardized features of particular devices, and external elements for communication of information external to the definition of the pictures in the CGM.

A minimal correct metafile consists of BEGIN METAFILE, a Metafile Descriptor consisting of METAFILE VERSION and METAFILE ELEMENT LIST, and END METAFILE.

4.2 Delimiter Elements

Every metafile starts with a BEGIN METAFILE element and ends with an END METAFILE element. This allows multiple metafiles to be stored or transferred together.

Each picture starts with a BEGIN PICTURE element and ends with an END PICTURE element. Between these delimiters, the Picture Descriptor is separated from the picture body by a BEGIN PICTURE BODY element.

Once the Metafile Descriptor has been read, access to individual pictures, on a random as opposed to sequential basis, may be safely accomplished if the encoding, access mechanism and implementation permit.

Delimiter Elements

BEGIN METAFILE and BEGIN PICTURE both have parameters for a name by which the m
picture (respectively) can be identified.

4.3 Metafile Descriptor Elements

The Metafile Descriptor (MD) is a group of elements that describes the functional capabiliti
to interpret the CGM. These elements are

METAFILE VERSION	MAXIMUM COLOUR INDEX
METAFILE DESCRIPTION	COLOUR VALUE EXTENT
VDC TYPE	METAFILE ELEMENT LIST
INTEGER PRECISION	METAFILE DEFAULTS
REAL PRECISION	REPLACEMENT
INDEX PRECISION	FONT LIST
COLOUR PRECISION	CHARACTER SET LIST
COLOUR INDEX PRECISION	CHARACTER CODING ANNOUNCER

In a particular metafile, the METAFILE ELEMENT LIST lists at least those standardize
that occur in the metafile. The CGM interpreter is thus informed of the capabilities required
fully interpret the Computer Graphics Metafile. The CGM contains a single Metafile Descr
Metafile Descriptor immediately follows the BEGIN METAFILE element in a metafile (with t
exception of intervening external and escape elements).

4.3.1 Identification

The identifying information includes declaration of the version of the CGM standard and
information about the origin, owner, generation date, etc., of the metafile.

4.3.2 Functional Capability

The contents of the Computer Graphics Metafile are defined by the METAFILE ELEMEN
ment. This contains a list of the control elements, graphical primitive elements, and attribu
that are utilized in the metafile. Two shorthand names for CGM elements are also provi
with the METAFILE ELEMENT LIST. The shorthand names shall not be considered macro
shall they be construed to be levels of conformance.

4.3.2.1 Drawing Set. The drawing set includes the mandatory CGM elements (i.e., those
appear in every conforming CGM) and most of the graphical primitive elements and attribut
The drawing set is specified by the shorthand name DRAWING SET.

The elements included in the drawing set are:

BEGIN METAFILE	LINE TYPE
END METAFILE	LINE WIDTH
BEGIN PICTURE	LINE COLOUR
BEGIN PICTURE BODY	MARKER BUNDLE INDEX
END PICTURE	MARKER TYPE
METAFILE VERSION	MARKER SIZE
METAFILE DESCRIPTION	MARKER COLOUR
VDC TYPE	TEXT BUNDLE INDEX
METAFILE ELEMENT LIST	TEXT FONT INDEX
AUXILIARY COLOUR	TEXT PRECISION
TRANSPARENCY	CHARACTER EXPANSION FACTO
CLIP RECTANGLE	CHARACTER SPACING
CLIP INDICATOR	TEXT COLOUR
VDC EXTENT	CHARACTER HEIGHT
BACKGROUND COLOUR	CHARACTER ORIENTATION

COLOUR SELECTION MODE   TEXT PATH
POLYLINE   TEXT ALIGNMENT
DISJOINT POLYLINE   FILL BUNDLE INDEX
POLYMARKER   INTERIOR STYLE
TEXT   FILL COLOUR
RESTRICTED TEXT   HATCH INDEX
APPEND TEXT   PATTERN INDEX
POLYGON   EDGE BUNDLE INDEX
POLYGON SET   EDGE TYPE
CELL ARRAY   EDGE WIDTH
GENERALIZED DRAWING PRIMITIVE   EDGE COLOUR
RECTANGLE   EDGE VISIBILITY
CIRCLE   FILL REFERENCE POINT
CIRCULAR ARC 3 POINT   PATTERN TABLE
CIRCULAR ARC 3 POINT CLOSE   PATTERN SIZE
CIRCULAR ARC CENTRE   COLOUR TABLE
CIRCULAR ARC CENTRE CLOSE   ASPECT SOURCE FLAGS
ELLIPSE   ESCAPE
ELLIPTICAL ARC   MESSAGE
ELLIPTICAL ARC CLOSE   APPLICATION DATA
LINE BUNDLE INDEX

4.3.2.2 Drawing Plus Control Set. The drawing-plus-control set may be used to indicate all of the elements in the drawing set plus additional control, Metafile Descriptor, Picture Descriptor, and attribute elements. It is specified by the shorthand name DRAWING PLUS CONTROL SET.

The elements included in the drawing-plus-control set are all of the elements in the drawing set and the following elements:

INTEGER PRECISION   CHARACTER CODING ANNOUNCER
REAL PRECISION   VDC INTEGER PRECISION
INDEX PRECISION   VDC REAL PRECISION
COLOUR PRECISION   SCALING MODE
COLOUR INDEX PRECISION   LINE WIDTH SPECIFICATION MODE
MAXIMUM COLOUR INDEX   MARKER SIZE SPECIFICATION MODE
COLOUR VALUE EXTENT   EDGE WIDTH SPECIFICATION MODE
METAFILE DEFAULTS REPLACEMENT   CHARACTER SET INDEX
FONT LIST   ALTERNATE CHARACTER SET INDEX
CHARACTER SET LIST

4.3.3 Default Metafile State

The default state is the state to which the interpreter is returned at the start of each picture. The default states of all metafile elements are defined in clause 6. These default values may be selectively replaced by using the METAFILE DEFAULTS REPLACEMENT element. The correspondence between character set indexes and registered or private character sets, and the meaning assigned to text font indexes, are also established in the Metafile Descriptor.

4.4 Picture Descriptor Elements

Picture Descriptor elements include elements to declare the parameter modes of other elements for an entire picture, to configure that portion of coordinate space that is of interest in the picture, and to set the colour to which the view surface is cleared at the start of the picture. These elements are SCAL-ING MODE, COLOUR SELECTION MODE, LINE WIDTH SPECIFICATION MODE, MARKER SIZE SPECIFICATION MODE, EDGE WIDTH SPECIFICATION MODE, VDC EXTENT, and BACK-GROUND COLOUR. If included in a picture, they shall appear after the BEGIN PICTURE element and before the BEGIN PICTURE BODY element. Escape and external elements are permitted in the

Picture Descriptor.

4.4.1 Scaling Mode

VDC space may be either an abstract space, which may be mapped to an arbitrary size on a physical device, or a metric space, which is intended to be mapped to a particular size. Selection of the mode to be used can be made on a picture-by-picture basis by means of the SCALING MODE element. The scaling mode element provides a flag to select abstract space or metric space, and a scale factor which specifies the number of millimeters per VDC unit when metric space is selected.

4.4.2 Colour Selection Mode

COLOUR SELECTION MODE selects either indexed or direct (RGB) colour specification for the picture and is described further under colour attributes.

4.4.3 Specification Modes

Line width, marker size, and edge width may be specified in more than one way. The width of lines, for example, may be specified as either a measure in VDC units or as a scale factor to be applied to a device-dependent nominal line width at interpretation time. For each attribute element having such multiple modes, there is an associated control element that defines the mode of the parameter of the attribute element.

4.4.4 VDC Extent

There is a metafile element to define the VDC extent. The extent is set with the VDC EXTENT element by specifying the addresses (in VDC) of the lower-left corner and the upper-right corner of this extent as seen by the viewer of the picture. Specification of values outside the VDC extent is permitted in CGM elements. It is intended that the visible portion of an image be contained within the VDC extent. It thus provides a frame for the region of interest in a picture. The values of the coordinates for either dimension may be either increasing or decreasing from the lower-left to the upper-right corner. For example, for devices with an upper-left origin, a picture may be described in coordinates that map directly to the device but still may be displayed correctly on a device with a lower-left origin. Figure 1 illustrates these concepts.

The VDC extent thus establishes the sense and orientation of VDC space (that is, the directions of the positive x (+x) and positive y (+y) axes, and whether the +y axis is 90-degrees clockwise or 90-degrees counterclockwise from the +x axis). In particular, VDC EXTENT establishes the direction of positive and negative angles as follows: positive 90-degrees is defined to be the right angle from the positive x-axis to the positive y-axis (see figure 1). Note that some attributes such as text attributes (for example, the directions of the 'up' and 'base' component vectors of CHARACTER ORIENTATION, and therefore the meaning of the enumerative values 'right', 'left', 'up', 'down') are intimately bound to these definitions.

The default state of the extent is specified in clause 6 and can be changed in the METAFILE DEFAULTS REPLACEMENT element in the MD. VDC EXTENT returns to this default state at the beginning of each picture.

4.4.5 CGM Tailoring

The ability to specify the VDC range and the VDC extent provides the flexibility to configure the metafile addressability in any way desired. It can be configured as an abstract, normalized address range for maximum device independence. It can also be configured to mimic the addressability of a particular target device in order to take advantage of particular device characteristics. The address range of such a device-specific metafile is just another normalized address range with the normalization limits inherent in the VDC-customizing element; therefore, device independence is maintained.

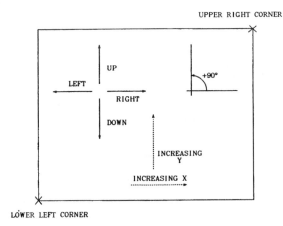

VDC EXTENT 0.0, 0.0, 1.0, 0.75

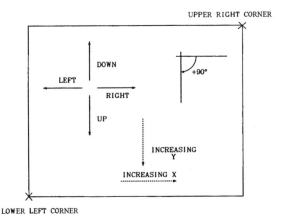

VDC EXTENT 0.0, 8.5, 11.0, 0.0

Figure 1. VDC EXTENT establishes the direction of positive and negative angles.

Such tailoring of the coordinates in a metafile can eliminate the need for transformation of coordinates at metafile interpretation time for the target device. The ability to specify the VDC extent thus allows for the exact registration of coordinates in a metafile with addressable points on the target graphics device.

ANSI X3.122 - 1986 Part 1   13

The use of VDC EXTENT to directly encode world coordinates of large dynamic range and very small granularity will likely result in performance penalties at metafile interpretation time, and may result in decreased portability if such VDC extents exceed those compatible with less capable metafile interpreters.

In addition to VDC tailoring, a metafile generator can limit or tailor the functional content of a metafile to accomodate particular devices or applications, and announce such functional tailoring through the use of METAFILE ELEMENT LIST.

4.4.6 Background Colour

Each picture defines a graphical image that is independent of the other images in a metafile. The background colour of the image may be specified by the BACKGROUND COLOUR Picture Descriptor element. If that element is not contained in the Picture Descriptor, the background colour of the image is the default background colour, whether that default is as specified in clause 6 or has been specified in the METAFILE DEFAULTS REPLACEMENT element.

The single parameter of BACKGROUND COLOUR is always RGB, regardless of the current value of COLOUR SELECTION MODE. If the COLOUR SELECTION MODE is indexed, then the BACKGROUND COLOUR element defines the initial representation of colour index 0 for the picture.

4.5 Control Elements

Control elements specify address space, clipping boundaries, and format descriptions of the CGM elements. Control of some of these format descriptions may be accomplished by Metafile Descriptor elements, while control of others is accomplished by control elements, which may appear in the picture bodies in the metafile. Those items in the former category are fixed for a given metafile, while those in the latter category are changeable; that is, they may change within a picture.

4.5.1 VDC Space and Range

The graphical primitive elements of a metafile define virtual images. The coordinates of these elements (that is, the addresses of points in the virtual image) are absolute two-dimensional Virtual Device Coordinates (VDC). VDC space is a two-dimensional coordinate space of infinite precision and infinite extent. Only a subset of VDC space, the VDC range, is realizable. The VDC range comprises all coordinates representable in the format specified by the declared VDC TYPE and (depending on the type) the VDC INTEGER PRECISION or VDC REAL PRECISION.

The VDC range is not directly settable; it is completely determined by VDC TYPE and either VDC INTEGER PRECISION or VDC REAL PRECISION elements in the metafile. These elements are controllable, some by dynamic elements in the metafile body and some by static elements in the MD. Note that the VDC range thus defined (a rectangular subregion of the VDC space) does not enclose a continuum of values, but has a distinct granularity. Regardless of the aspect ratio of the VDC range and the granularity within the range, it is implicit that one VDC unit in the x-direction represents the same distance as one VDC unit in the y-direction in VDC space.

4.5.2 Clipping

In order to defer clipping of graphical primitive elements (particularly, expandable elements such as CIRCLE, CIRCULAR ARC 3 POINT, TEXT, etc.) until metafile interpretation time, a clipping control feature is provided in the CGM. Clipping control is achieved by defining CLIP RECTANGLE in VDC space. Whether clipping to the limits of CLIP RECTANGLE actually occurs at metafile interpretation time is controlled by the CLIP INDICATOR element that sets the mode of the metafile to 'on' or 'off'. The defaults for CLIP RECTANGLE and CLIP INDICATOR are listed in clause 6.

raphical Primitive Elements

iical primitive elements are those elements that describe the visual components of a picture. Their
nate arguments are specified in VDC units. The CGM provides the graphical primitive elements

    POLYLINE
    DISJOINT POLYLINE
    POLYMARKER
    TEXT
    RESTRICTED TEXT
    APPEND TEXT
    POLYGON
    POLYGON SET
    CELL ARRAY
    GENERALIZED DRAWING PRIMITIVE (GDP)
    RECTANGLE
    CIRCLE
    CIRCULAR ARC 3 POINT
    CIRCULAR ARC 3 POINT CLOSE
    CIRCULAR ARC CENTRE
    CIRCULAR ARC CENTRE CLOSE
    ELLIPSE
    ELLIPTICAL ARC
    ELLIPTICAL ARC CLOSE.

metafile supports access to special geometric output capabilities of devices and workstations
igh the GDP. The GDP has a list of points in VDC as a parameter. It is thus well suited for non-
lardized output primitives, which have position, shape, extent, etc., whereas ESCAPE is better
1 for non-standardized device control functions.

'ormal definition of the CGM describes graphical primitive elements which are positionally indepen-
by virtue of containing complete .explicit positional information within each element definition.
"CP-less" model (without the concept of a Current Position) corresponds to that of GKS, and
is in fewer side effects and device interaction problems than the CP-oriented graphical models. The
ency advantages often associated with a CP-oriented model have been realized through techniques
in some of the encodings of this Standard.

TEXT, RESTRICTED TEXT, and APPEND TEXT elements and related text attribute elements
efined in the current VDC space. Thus, they are affected by changes to the Virtual Device Coordi-
format.

types or categories of graphical primitive elements are defined for the CGM: line elements, marker
int, text elements, filled-area elements and cell array element.

ine elements are:

    POLYLINE
    DISJOINT POLYLINE
    CIRCULAR ARC 3 POINT
    CIRCULAR ARC CENTRE
    ELLIPTICAL ARC

marker element is:

    POLYMARKER

The text elements are:

TEXT
RESTRICTED TEXT
APPEND TEXT

The filled-area elements are:

POLYGON
POLYGON SET
RECTANGLE
CIRCLE
CIRCULAR ARC 3 POINT CLOSE
CIRCULAR ARC CENTRE CLOSE
ELLIPSE
ELLIPTICAL ARC CLOSE

The cell array element is:

CELL ARRAY

In addition to these five classes of element, the GENERALIZED DRAWING PRIMITIVE (GDP) is a graphical primitive element that may be be used to access device (or implementation) specific graphical primitives that are not accessed by the standardized elements.

4.6.1 Line Elements

4.6.1.1 Description. There are two general line elements - POLYLINE and DISJOINT POLYLINE - and three line elements relating to circles and ellipses.

POLYLINE	generates a set of connected lines as defined by a list of points, starting with the first, drawing a line through each successive point, ending at the last point.
DISJOINT POLYLINE	generates a set of unconnected lines as defined by a list of point pairs, drawing from the first to the second, the third to the fourth, etc..
CIRCULAR ARC xxx	generates a single circular arc; two parameterizations of the arc are possible; these are described in 5.6.13 and 5.6.15.
ELLIPTICAL ARC	generates a single elliptical arc; the parameterization of the arc is described in 5.6.18.

4.6.1.2 Attributes. The appearance of all line elements is controlled by the line attributes, the LINE BUNDLE INDEX and the ASPECT SOURCE FLAGs associated with the line attributes that may be bundled. These are described in 4.7.1.

4.6.1.3 Usage of Line Elements. POLYLINE is the most general of the primitives. DISJOINT POLYLINE is intended for situations where the alternative would be a large number of "2 point" POLYLINE elements. The ARC primitives provide data compression by comparison with POLYLINE and allow the arcs to be described without knowledge of the resolution of the final viewing surface.

4.6.2 Marker Element

4.6.2.1 Description. There is a single marker element.

POLYMARKER	generates symbols of a specific type at each of a list of points.

4.6.2.2 Attributes. The appearance of the markers is controlled by the marker attributes, the MARKER BUNDLE INDEX and the ASPECT SOURCE FLAGs associated with the marker attributes that may be bundled. These are described in 4.7.2.

4.6.2.3 Usage of the Marker Element. Markers conceptually indicate the location of their specifying points. Therefore, if the CLIP INDICATOR is 'on', the marker is visible if, and only if, its specifying point is within the CLIP RECTANGLE. If its specifying point is inside the CLIP RECTANGLE but part of the marker lies outside, the manner in which the marker is clipped (or not) is not standardized.

In situations where there is a requirement that includes making visible those parts of a symbol that are inside the CLIP RECTANGLE while the specifying position of the symbol is outside, the most appropriate primitive element is the TEXT element, used with a single character text argument. For TEXT, three PRECISIONs are specified which control the precision with which each element is clipped. STROKE precision requires clipping even within the body of a symbol. Achieving the centring of the symbol at the specifying position is possible with the TEXT ALIGNMENT element.

4.6.3 Text Elements

4.6.3.1 Description. Three text elements are provided.

TEXT	generates a text string (or part of a text string) aligned to a particular point.
RESTRICTED TEXT	generates a text string (or part of a text string) that is constrained within a given area.
APPEND TEXT	generates a part of a text string as a part of a text string started with a TEXT or RESTRICTED TEXT element.

4.6.3.2 Attributes. The appearance of all text elements is controlled by the text attributes, the TEXT BUNDLE INDEX and the ASPECT SOURCE FLAGs associated with the text attributes that may be bundled. These are described in 4.7.3.

Changes to the text attributes TEXT FONT INDEX, CHARACTER EXPANSION FACTOR, CHARACTER SPACING, TEXT COLOUR, CHARACTER HEIGHT, CHARACTER SET INDEX, ALTERNATE CHARACTER SET INDEX, and TEXT BUNDLE INDEX, and to the control elements AUXILIARY COLOUR and TRANSPARENCY, are permitted between a non-final text element and its succeeding APPEND TEXT element.

4.6.3.3 Usage of Text Elements. Each text element has a 'final/not-final' flag. This permits a text string to be started with a TEXT or RESTRICTED TEXT element and continued with one or more APPEND TEXT elements. Only the last element will have its flag set to 'final'. The initial element is always TEXT or RESTRICTED TEXT; subsequent elements may only be APPEND TEXT.

The attributes that may be changed between a related set of text elements are listed in the description of APPEND TEXT (see 5.6.6). These include those that affect font and character set changes, character size and text colour.

The current setting of TEXT ALIGNMENT is used to align the complete text string assembled from the separate text elements.

4.6.4 Filled-Area Elements

4.6.4.1 Description. There are two general fill elements: POLYGON and POLYGON SET. In addition there are six elements that provide data compression and allow areas to be filled accurately without knowledge of the resolution of the final viewing surface.

POLYGON	generates an area and its edge, defined by a list of points; the style of the area is one of 'hollow', 'solid', 'pattern', 'hatch', or 'empty'; the visibility and style of the edge of the area depend on the edge attributes alone.
POLYGON SET	generates a number of areas and their edges, defined by a list of vertex points and vertex flags; the set of styles is the same as for POLYGON; the vertex flags indicate the different polygons in the set;

the vertex flags and the edge attributes together control the visibility and style of individual edge segments of each polygon.

RECTANGLE   generates an upright rectangular area; the set of styles is the same as for POLYGON.

CIRCLE   generates a circle; the set of styles is the same as for POLYGON.

CIRCULAR ARC xxx CLOSE   generates a partial circular area; 'pie' and 'chord' style arcs are possible; two parameterization of the arcs are provided; these are described in 5.6.14 and 5.6.16; the set of styles is the same as for POLYGON.

ELLIPSE   generates an ellipse; the parameterization of the ellipse is described in 5.6.17; the set of styles is the same as for POLYGON.

ELLIPTICAL ARC CLOSE   generates a partial elliptical area; 'pie' and 'chord' style arcs are possible; the parameterization is described in 5.6.19; the set of styles is the same as for POLYGON.

4.6.4.2 Attributes. The appearance of all filled-area elements is controlled by the fill attributes, the FILL BUNDLE INDEX, the EDGE BUNDLE INDEX, and the ASPECT SOURCE FLAGs associated with the fill attributes that may be bundled. These are described in 4.7.4.

4.6.4.3 Usage of Fill Elements. POLYGON provides for the representation of standard irregular areas. RECTANGLE, because it is upright, is a more efficient parameterization of a rectangle than a POLYGON and may be implemented directly in some systems.

The circular and elliptical fill primitives provide an efficient parameterization and allow the areas to be produced accurately without knowledge of the resolution of the final viewing surface.

POLYGON SET allows a related set of polygons to be represented. All attributes of each of the polygons are the same. The specification of the vertex flags allows disjoint polygons (such as both the body and the dot of the letter 'i'), holes (as in a broad ring) and overlapping areas. Accurate rendering of abutting areas of uniform colour, pattern or hatch is possible with the control provided over individual edge visibility.

4.6.4.4 Interior. The interior of a filled-area element is defined as follows. For a given point, create a straight line starting at that point and going to infinity. If the number of intersections between the straight line and the filled area is odd, the point is within the filled area; otherwise it is outside. If the straight line passes a filled-area vertex tangentially, the intersection count is not affected. If a point is within the filled area, it is included in the area to be filled subject to the rules for boundaries and edges (see 4.7.8).

4.6.4.5 Clipping. If parts of a filled-area element are clipped, then the intersection of the interior and the clip boundary becomes part of the boundary of the resulting clipped area for the purposes of display of the boundary for interior style 'hollow'. If the edge is visible, it is not drawn along the new boundary segments created by the clipping of the area.

4.6.5 Cell Array Element

CELL ARRAY   represents a 2-dimensional array of colour values, which cover a rectangle or parallelogram.

The colour values are either direct colour values or indexes into the COLOUR TABLE, according to the current COLOUR SELECTION MODE. The colour values are in the precision declared by a 'local colour precision' parameter of the CELL ARRAY element.

CELL ARRAY is not controlled by any attributes.

4.6.6 Circular Arc Elements

The CGM provides for two forms of specification of circular arc elements: a centre-radius specification and a 3-point specification. Each has its advantages and disadvantages with respect to numerical accuracy, relationship of defining data to the VDC range, etc.

When choosing which parameterization to use, one should decide where possible numerical inaccuracy would be least disturbing. The 3-point form specifies exact arc endpoints, but might result in inaccurate centre-point calculations, whereas the centre form specifies exact centre point but might result in roundoff errors on the ends of the arc. The 3-point form would thus be more appropriate for smoothly joining an arc to a polyline in a line drawing, whereas the centre form would be more appropriate for pie charts.

4.6.7 Elliptical Elements

4.6.7.1 Geometric Concepts. Ellipses are specified by Conjugate Diameter Pairs. A Conjugate Diameter Pair (CDP) of an ellipse is a pair D,d of diameters of the ellipse such that a tangent to the ellipse at each endpoint is parallel to the other diameter. The four tangents to the ellipse at the endpoint of the CDP thus form a parallelogram whose sides are bisected by the endpoints of the diameters.

Any CDP of the ellipse remains a CDP across any graphical transformation which transforms an ellipse into an ellipse. This is demonstrated in figure 2 in which the ellipse has been scaled by a factor of two in the y-direction only.

Thus any CDP of a desired ellipse can be used to specify the ellipse. Note that the (mutually perpendicular) major and minor axes of an ellipse and any pair of perpendicular diameters of a circle are CDP's, although they do not necessarily remain perpendicular across a transformation.

Thus to specify an ellipse, all that is needed is three points:

— the centrepoint of the ellipse;

— two CDP endpoints (one endpoint from each diameter).

4.6.7.2 Parameterization of Elliptical Elements in CGM. The ellipse itself in each of the three elliptical elements is parameterized as in the preceding section, the centrepoint and two CDP endpoints. For the two elliptical arc elements, the start and end of the defined arc section is parameterized by two semi-infinite rays originating at the centrepoint. The intersection of these rays with the ellipse defines two points on the ellipse, and these two points define the arc.

4.7 Attribute Elements

Attribute elements determine the appearance of graphical primitive elements. Attributes are classified as either individual attributes or attributes that may be bundled. Table 1 lists the attributes by this classification.

Bundled selection of attributes implies that the appearances of graphical primitive elements are distinguishable from one another when different bundles are specified. The method of specification of the aspects that may be bundled of a graphical primitive element may be chosen separately for each aspect. A further group of attributes called ASPECT SOURCE FLAGS (ASFs) takes the values 'individual' and 'bundled' to specify the choice. There is one ASF for each aspect that may be bundled of each primitive.

There is a current modal value for every attribute. Elements are provided to change these modal values. The modal value established by setting an attribute remains until it is explicitly changed. All attributes return to their default values when the BEGIN PICTURE element is encountered.

There is at least one bundle index associated with each of the graphical primitive element types — line, marker, filled area, and text. Line, marker, and text elements have a single associated bundle index. Filled-area elements have two associated bundle indexes, one for interior attributes and one for edge

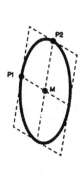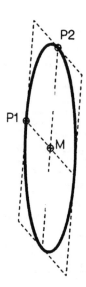

Figure 2. Anisotropic scaling of an ellipse.

attributes.

The value of each bundle index attribute is modally bound to subsequent graphical primitive elements of the associated type. Distinct values of the bundle index correspond to distinct appearances of the graphical primitive element. For each attribute that may be bundled, there is an associated Aspect Source Flag (ASF).

For individual attributes, the current modal value is used to display a graphical primitive element. For attributes that may be bundled a graphical primitive element is displayed as follows:

TABLE 1. Individual Attributes and Attributes that May Be Bundled

Individual	May Be Bundled
CHARACTER HEIGHT	LINE TYPE
CHARACTER ORIENTATION	LINE WIDTH
TEXT PATH	LINE COLOUR
TEXT ALIGNMENT	MARKER TYPE
CHARACTER SET INDEX	MARKER SIZE
ALTERNATE CHARACTER SET INDEX	MARKER COLOUR
EDGE VISIBILITY	TEXT FONT INDEX
FILL REFERENCE POINT	TEXT PRECISION
PATTERN SIZE	CHARACTER EXPANSION FACTOR
	CHARACTER SPACING
	TEXT COLOUR
	INTERIOR STYLE
	FILL COLOUR
	HATCH INDEX
	PATTERN INDEX
	EDGE TYPE
	EDGE WIDTH
	EDGE COLOUR

a) if the ASF for an aspect is 'individual', the value used is the current modal value (which is set only by the individual aspect-setting elements);

b) if the ASF for an aspect is 'bundled', the value used is obtained via the bundle table for that primitive; the corresponding component of the bundle, which is pointed to by the bundle index, is used.

The actual resulting appearance is interpreter dependent, but the intent is that the interpreter render distinct appearances of graphical primitive elements for distinct values of the associated bundle index (or indexes) by manipulation of the attributes that may be bundled. For example, LINE BUNDLE INDEX designates visually distinct combinations of the polyline attributes LINE WIDTH, LINE TYPE, and LINE COLOUR. Table 2 lists the aspects of each bundle.

Because inquiry of bundle representations is not generally possible in a metafile environment, mixing of 'individual' and 'bundled' ASF values within a bundle will compromise the guarantee of distinguishability of different bundle indexes within that bundle at interpretation time.

4.7.1 LINE Bundle

The LINE BUNDLE INDEX selects one entry in a table of bundled attribute values. The following attributes are in this bundle:

a) LINE TYPE: determines the type of the line (for example, 'dotted', 'dashed', etc.) with which the polyline is rendered;

b) LINE WIDTH: determines the width of the line with which the polyline is rendered;

c) LINE COLOUR: determines the colour in which the polyline is drawn.

4.7.2 MARKER Bundle

The MARKER BUNDLE INDEX selects one entry in a table of bundled attribute values. The following attributes are in this bundle:

a) MARKER TYPE: determines the symbol that is drawn at the marker position (for example, 'dot', 'plus', etc.);

ANSI X3.122 - 1986 Part 1   21

TABLE 2. Aspects of the Bundle and Affected Primitives

Bundle	Aspects	Affected Primitives
LINE	LINE TYPE LINE WIDTH LINE COLOUR	POLYLINE DISJOINT POLYLINE CIRCULAR ARC 3 POINT CIRCULAR ARC CENTRE ELLIPTICAL ARC
MARKER	MARKER TYPE MARKER SIZE MARKER COLOUR	POLYMARKER
FILL	INTERIOR STYLE FILL COLOUR HATCH INDEX PATTERN INDEX	POLYGON POLYGON SET RECTANGLE CIRCLE CIRCULAR ARC 3 POINT CLOSE CIRCULAR ARC CENTRE CLOSE ELLIPSE ELLIPTICAL ARC CLOSE
EDGE	EDGE TYPE EDGE WIDTH EDGE COLOUR	POLYGON POLYGON SET RECTANGLE CIRCLE CIRCULAR ARC 3 POINT CLOSE CIRCULAR ARC CENTRE CLOSE ELLIPSE ELLIPTICAL ARC CLOSE
TEXT	TEXT FONT INDEX TEXT PRECISION CHARACTER EXPANSION FACTOR CHARACTER SPACING TEXT COLOUR	TEXT RESTRICTED TEXT APPEND TEXT

b) MARKER SIZE: determines the size of the marker symbol;

c) MARKER COLOUR: determines the colour in which the marker symbol is drawn.

4.7.3 TEXT Bundle

The TEXT BUNDLE INDEX selects one entry in a table of bundled attribute values. The following attributes are in this bundle:

a) TEXT FONT INDEX: determines the style of the graphical display of the text characters;

b) TEXT PRECISION: determines the fidelity with which characters need be displayed and positioned;

c) CHARACTER EXPANSION FACTOR: determines the deviation of the character width/height ratio from the ratio established by the font designer;

d) CHARACTER SPACING: determines the amount of blank space added between characters in a string;

e) TEXT COLOUR: determines the colour in which the text characters are drawn.

4.7.4 Filled-area bundles

There are two bundles associated with filled-area elements.

4.7.4.1 FILL bundle. The FILL bundle is associated with the interior attributes of filled-area elements. The FILL BUNDLE INDEX selects one entry in a table of bundled attribute values. The following attributes are in this bundle:

a) INTERIOR STYLE: determines which of the classes of interior ('hollow', 'solid', 'pattern', 'hatch', or 'empty') is used to draw a filled-area element;

b) FILL COLOUR: determines the colour in which the interior of a filled-area primitive is drawn. This applies only to interior styles 'hollow', 'solid' and 'hatch' (the drawn boundary of a 'hollow' area is considered as part of the representation of the interior);

c) HATCH INDEX: determines which hatch style is used if 'hatch' interior style is selected;

d) PATTERN INDEX: determines which entry in the pattern table is used if 'pattern' interior style is selected.

4.7.4.2 EDGE bundle. The EDGE bundle is associated with the edge attributes of filled-area elements. The EDGE BUNDLE INDEX selects one entry in a table of bundled attribute values. The following attributes are in this bundle:

a) EDGE TYPE: determines the line type with which the edges are drawn;

b) EDGE WIDTH: determines the width of the edge;

c) EDGE COLOUR: determines the colour in which the edge is drawn.

4.7.5 Specification Modes

The CGM provides the mechanism for both 'absolute' and 'scaled' specification of the modal values of the size-related elements LINE WIDTH, MARKER SIZE, and EDGE WIDTH. 'Absolute' specification means that the sizes are given in VDC units. 'Scaled' specification means that the size is specified as a scale factor to be applied at metafile interpretation time to the device-dependent nominal size for the associated primitive.

4.7.6 TEXT Attributes

The representation and placement of text characters on a device is controlled by the attribute elements TEXT FONT INDEX, CHARACTER SET INDEX, ALTERNATE CHARACTER SET INDEX, TEXT PRECISION, CHARACTER EXPANSION FACTOR, CHARACTER SPACING, TEXT COLOUR, and CHARACTER HEIGHT and by the control elements AUXILIARY COLOUR and TRANSPARENCY. The placement and orientation of text strings is controlled by the attribute elements CHARACTER ORIENTATION, TEXT PATH, and TEXT ALIGNMENT. TEXT BUNDLE INDEX is an index into the text bundle table, each entry of which contains values for the attributes that may be bundled. Although the placement and size of text can be precisely specified by the attributes mentioned, the fidelity of rendering depends on the current TEXT PRECISION.

The choice of character font (that is, the style of the characters to be displayed) is determined independently of the character set. However, the specified font will only have meaning if it is related to the character set being used. Roman and Gothic are examples of commonly used fonts for Latin-based alphabets.

The attributes in the character representation and placement group (above) and TEXT BUNDLE INDEX may be changed within a string. A TEXT element or RESTRICTED TEXT element is tagged to show it is not complete and provides only the first portion of the string. The TEXT element or RESTRICTED TEXT element may be followed by the desired text attribute element(s) and then by an APPEND TEXT element, which provides the next portion of the string. This may be repeated as often as necessary, with the final APPEND TEXT tagged to indicate that the string is complete. Note that a metafile interpreter generally cannot display any of the text until the string is complete because of TEXT ALIGNMENT and the way in which attribute changes affect the definition of the text extent rectangle (see below). Text may be displayed before the string is complete only in the following cases:

Path	Vertical Alignment	Horizontal Alignment
right	normal vertical or baseline	normal horizontal, left, or continuous (0,0)
left	normal vertical or baseline	normal horizontal, right, or continuous (1,0)
down	top, capline, normal vertical, or continuous (0,1)	normal horizontal or centre
up	baseline, bottom, normal vertical, or continuous (0,0)	normal horizontal or centre

There are several methods for inclusion within a string of characters from different character sets. The method used is determined by the CHARACTER CODING ANNOUNCER Metafile Descriptor element. The default or normal technique is to use the CHARACTER SET INDEX element, and restrict the contents of the text strings to printing characters and spaces (format effector control codes such as CR and LF are permitted, but their interpretation is implementation-dependent). Other settings of the CHARACTER CODING ANNOUNCER or use of the ALTERNATE CHARACTER SET INDEX element permit standardized use of 8-bit characters and the SI, SO, and ESC control codes within the text string, in accordance with ISO 2022. The ALTERNATE CHARACTER SET INDEX element is used to select a character set to be used as both the G1 set and the G2 set. The G1 set is used both for 8-bit characters in columns 10-15 of the code table, and with the SO control code. The assignment of meaning to the index parameter of both CHARACTER SET INDEX and ALTERNATE CHARACTER SET INDEX is done with the CHARACTER SET LIST Metafile Descriptor element.

Selection of fonts from different font tables is done by the TEXT FONT INDEX element. The assignment of meaning to the index values of TEXT FONT INDEX is done with the Metafile Descriptor element FONT LIST.

The font coordinate system is illustrated in figure 3. The character body encloses all of the drawn parts (kerning excepted) of all characters in the font (that is, no descender extends lower than 'bottom', and no accent mark or oversized symbol extends higher than 'top'). The left and right edges of the character body may be defined on a per-character basis to accommodate variable widths, and proportional spacing. It is expected that font designers will specify some fonts having kerns extending beyond the character body. The body exceeds the actual character symbol width and height as necessary to provide adequate white space between characters, such that text is readable and adequately separated when adjacent character bodies are flush (that is, when CHARACTER SPACING is 0). The character body is defined in this way to permit alignment of multiline text without overlaps in the metafile environment. The CHARACTER HEIGHT specifies the VDC distance between the capline and baseline of the font (see figure 3). The CHARACTER EXPANSION FACTOR specifies the deviation of the width to height ratio of the characters from the ratio indicated by the font designer (see figure 4). CHARACTER SPACING specifies how much additional space is to be inserted between two adjacent character bodies (see figure 5). If the value of CHARACTER SPACING is zero, the character bodies are arranged one after the other along the TEXT PATH with only the intercharacter spacing designated by the font designer. If the value of CHARACTER SPACING is positive, additional space is inserted between character bodies. If the value of CHARACTER SPACING is negative, adjacent character bodies overlap although the character symbols themselves might not. Character spacing is specified as a fraction of the CHARACTER HEIGHT.

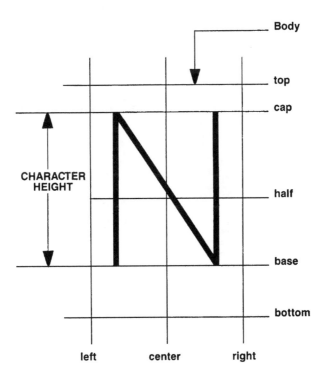

Figure 3. Font description coordinate system.

 **CHARACTER HEIGHT = 1.0**
CHARACTER EXPANSION FACTOR = 1.0

 **CHARACTER HEIGHT = 1.5**
CHARACTER EXPANSION FACTOR = 1.0

 **CHARACTER HEIGHT = 1.0**
CHARACTER EXPANSION FACTOR = 1.5

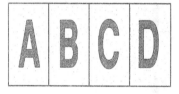 **CHARACTER HEIGHT = 2.0**
CHARACTER EXPANSION FACTOR = 0.75

Figure 4. CHARACTER HEIGHT and CHARACTER EXPANSION FACTOR.

CHARACTER HEIGHT = 1.0
CHARACTER SPACING = 0.67
TEXT PATH = right

CHARACTER HEIGHT = 1.0
CHARACTER SPACING = −0.67
TEXT PATH = right

CHARACTER HEIGHT = 1.0
CHARACTER SPACING = 2.0
TEXT PATH = down

Figure 5. CHARACTER SPACING.

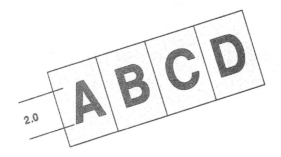

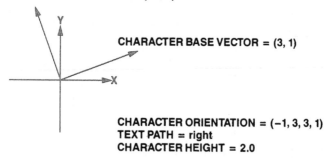

CHARACTER UP VECTOR = (−1, 3)

CHARACTER BASE VECTOR = (3, 1)

CHARACTER ORIENTATION = (−1, 3, 3, 1)
TEXT PATH = right
CHARACTER HEIGHT = 2.0

Figure 6. CHARACTER ORIENTATION.

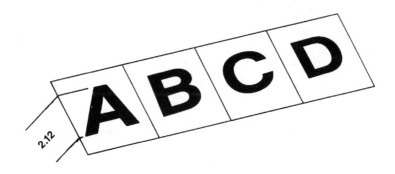

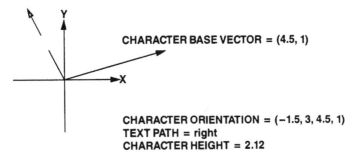

CHARACTER UP VECTOR = (−1.5, 3)

CHARACTER BASE VECTOR = (4.5, 1)

CHARACTER ORIENTATION = (−1.5, 3, 4.5, 1)
TEXT PATH = right
CHARACTER HEIGHT = 2.12

Figure 7. CHARACTER HEIGHT and CHARACTER ORIENTATION after anisotropic transformation.

CHARACTER ORIENTATION specifies the character up vector and base vector, which fix the orientation, skew, and distortion of the characters, and also determine the sense of 'right', 'left', 'up', and 'down' for TEXT PATH and TEXT ALIGNMENT (see figure 6).

The way in which software above the metafile generator and/or the metafile generator itself may use CHARACTER ORIENTATION is described. To generate the CHARACTER ORIENTATION and CHARACTER HEIGHT elements, a vector whose length is the character height (baseline-to-capline) and whose direction is the desired character up vector is created. A second vector is also created with the same length, whose direction is negative 90-degrees from the up vector. This pair of vectors may be transformed before being given to the metafile generator as the parameters to CHARACTER ORIENTATION. The length of the transformed up vector may then be used to generate the CHARACTER HEIGHT element. If an anisotropic transformation is in effect above the metafile generator, the character height must be respecified by the metafile generator for each change in orientation (see figure 7). The CHARACTER HEIGHT and CHARACTER ORIENTATION are decoupled to permit changing character height (but not orientation) within a string. Thus, to the metafile interpreter, the absolute lengths of the vectors in CHARACTER ORIENTATION are not significant; only their directions and the ratio of their lengths are significant.

The ratio of the length of the width vector to the length of the height vector is used to scale the CHARACTER SPACING for text paths 'right' and 'left', and the CHARACTER EXPANSION FACTOR in all cases, before these are used to display the text.

TEXT PATH has the possible values 'right', 'left', 'up', and 'down'. It specifies the writing direction of the text string as follows:

right: means the direction of the character base vector;

left: means 180-degrees from the character base vector;

up: means the direction of the character up vector;

down: means 180-degrees from the character up vector.

For the 'up' and 'down' text path directions, the characters are arranged so that the centres of the character bodies are on a straight line in the direction of the up vector of CHARACTER ORIENTATION. For the 'left' and 'right' text path directions, the characters are arranged so that the baselines of the characters are on a straight line parallel to the direction of the character base vector. These composition rules also hold true when characters of different heights, expansion factors, fonts, or precisions are intermixed in a string by means of attribute changes between non-final TEXT elements and subsequent APPEND TEXT elements.

Alignment of text is done with respect to a text extent rectangle, which is derived by joining the character bodies of the characters in the string according to the current status of the attributes and the composition rules described. Alignment is performed according to the highest precision in the string.

For TEXT PATH = 'left' or 'right',

TOPLINE:	topline farthest from the baseline
CAPLINE:	capline farthest from the baseline
HALFLINE:	halfline farthest from the baseline
BOTTOMLINE:	bottomline farthest from the baseline
LEFT:	leftmost edge of leftmost character body
RIGHT:	rightmost edge of rightmost character body
CENTRE:	halfway between left and right edges

For TEXT PATH = 'up' or 'down',

TOPLINE:	topline of topmost character
CAPLINE:	capline of topmost character
HALFLINE:	halfway between halflines of topmost and bottommost character
BASELINE:	baseline of bottommost character
BOTTOMLINE:	bottomline of bottommost character

LEFT: left edge farthest from the centreline
RIGHT: right edge farthest from the centreline

Note that the relationship of topline to capline, bottomline to baseline, and the placement of the halfline are font-dependent (see figure 8). It is for this reason that the various defining lines of the text extent rectangle need not be derived from the same character body. This is a function of the text height, text font, text precision and character expansion factor changes within a string.

The TEXT ALIGNMENT attribute controls the positioning of the text extent rectangle in relation to the text position (see figure 9).

The horizontal component of TEXT ALIGNMENT has five possible values: 'left', 'centre', 'right', 'normal horizontal' and 'continuous horizontal'. If the horizontal component is 'left', the left side of the text extent rectangle passes through the text position. Similarly, if the value is 'right', the right side of the text extent rectangle passes through the text position. If the horizontal component is 'centre', the text position lies midway between the left and right sides of the text extent rectangle. In this case, if TEXT PATH = 'up' or 'down', the straight line passing through the centrelines of the characters also passes through the text position.

The vertical component of TEXT ALIGNMENT has seven possible values: 'top', 'cap', 'half', 'base', 'bottom', 'normal vertical', and 'continuous vertical'. A vertical alignment value of 'top', 'cap', 'half', 'base', or 'bottom' causes the text to be moved such that the corresponding defining line of the text extent rectangle passes through the text position.

For both horizontal and vertical alignment, normal values are converted to the appropriate value, as indicated in clause 5, at text element elaboration time and thereafter treated as above. For all values of TEXT ALIGNMENT, the alignment value applies to the complete text string, which may be comprised of non-final partial strings and a final partial string.

If the value of the horizontal component of TEXT ALIGNMENT is 'continuous horizontal', an additional value, 'continuous horizontal alignment' (a real number normalized so that 1.0 corresponds to the width of the text extent rectangle) is used as an offset from the text position to the left side of the text extent rectangle. Figure 10 illustrates the sense of positive and negative values of 'continuous horizontal alignment'.

If the value of the vertical component of TEXT ALIGNMENT is 'continuous vertical', an additional value, 'continuous vertical alignment' (a real number normalized so that 1.0 corresponds to the height of the text extent rectangle) is used as an offset from the text position to the bottom side of the text extent rectangle. Figure 10 illustrates the sense of positive and negative values of 'continuous vertical alignment'.

The foregoing examples have been illustrated for the case of the character up vector and the character base vector being orthogonal. When they are not, the text extent rectangle becomes a parallelogram, with the sides remaining parallel to the two orientation vectors. The centreline skews to remain parallel with the left and right edges of the text extent parallelogram. The height of the text extent rectangle is measured along the skewed edge (not perpendicular to the baseline), and the distance to be moved for alignment is done along the angle made by the appropriate orientation vector (see figure 11). Right is in the direction of the character base vector, and left is in the opposite direction.

TEXT ALIGNMENT = (centre, cap, 0, 0)

TEXT PATH = right
CHARACTER HEIGHT = 2.0
String = Big
CHARACTER HEIGHT = 1.0
Appended String = Little

TEXT ALIGNMENT = (right, half, 0, 0)

TEXT PATH = down
CHARACTER HEIGHT = 1.0
CHARACTER EXPANSION FACTOR = 1.0
String = Normal
CHARACTER EXPANSION FACTOR = 2.0
Appended String = Wide

Figure 8. Discrete text alignment with appended text and proportional spacing.

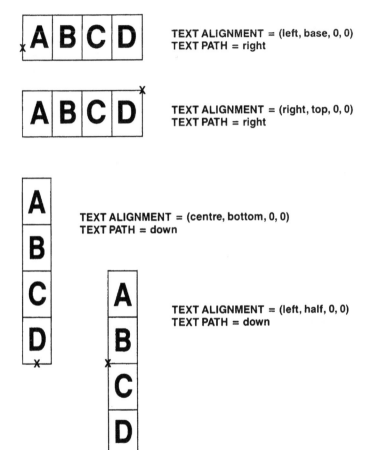

Figure 9. Discrete text alignment.

TEXT ALIGNMENT = (continous horizontal, base, 0.25,
TEXT PATH = right

TEXT ALIGNMENT = (continuous horizontal, continuou
vertical, −0.25, −0.25)
TEXT PATH = right

TEXT ALIGNMENT = (left, continuous vertical, 0, 1)
TEXT PATH = right
String 1 = ABCD

TEXT ALIGNMENT = (left, continuous vertical, 0, 2.5)
TEXT PATH = right
String 2 = EFGH

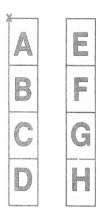

TEXT ALIGNMENT = (continuous horizontal, top, 0, 0)
TEXT PATH = down
String 1 = ABCD

TEXT ALIGNMENT = (continuous horizontal, top, 2.0, 0)
TEXT PATH = down
String 2 = EFGH

Figure 10. Continuous text alignment.

X

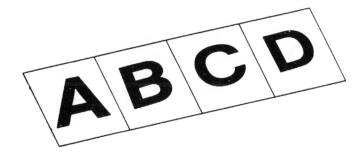

CHARACTER UP VECTOR = (–1.5, 3)

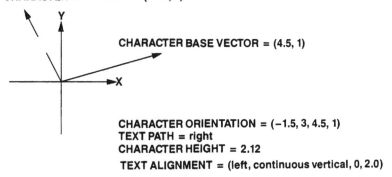

CHARACTER BASE VECTOR = (4.5, 1)

CHARACTER ORIENTATION = (–1.5, 3, 4.5, 1)
TEXT PATH = right
CHARACTER HEIGHT = 2.12
TEXT ALIGNMENT = (left, continuous vertical, 0, 2.0)

Figure 11. Continuous text alignment after anisotropic transformation.

The continuous values of alignment allow proper positioning of multiple rows or columns of text relative to each other, using a single text position for all of the rows or columns. Rows might typically consist of several lines from a horizontally written character set displayed with 'right' path. Columns might consist of strings of a vertically written alphabet displayed with 'down' path.

Such positioning would not otherwise be achievable in a metafile environment, because inquiry of the dimensions of the text extent rectangle cannot be provided at metafile generation time.

As an example of the set of the continuous alignment attributes, consider the display of four rows of left-justified text, each consisting of a single string specified by a single TEXT element. To ensure that ascenders and descenders do not interfere between rows and that, in addition, there is a space of at least one-half the maximum size of a character between the descenders of one row with raised accent marks or oversized symbols of another row, TEXT ALIGNMENT should be set to ('left', 'continuous vertical'), and 'vertical alignment' to 0.0. Then, output the first row with the text position equal to the lower-left corner of the string. Now, set 'vertical alignment' to 1.5, and output the second string with the same text position. This places the second row below the first because of the change in alignment. The last two rows are output in the same way with 'vertical alignment' set to 3.0 and 4.5, and the text position parameters to TEXT unchanged. A value of 1.0 assures no overlap between rows; anything greater than 1.0 guarantees additional non-printing space.

TEXT PRECISION is used to specify the 'closeness' of the text representation at metafile interpretation in relation to that defined by the other metafile text attributes and the clipping currently applicable. The following precision values are defined:

string: The complete text string is generated in the requested text font and is positioned by aligning the string at the given text position. Text height and CHARACTER EXPANSION FACTOR are evaluate as closely as possible given the capabilities of the metafile interpreter. The text vectors, TEXT PATH, TEXT ALIGNMENT, and CHARACTER SPACING need not be used. Clipping is done in an implementation dependent fashion.

char: The complete text string is generated in the requested text font. For the representation of each individual character, the aspects text height, the text vectors, and CHARACTER EXPANSION FACTOR are evaluated as closely as possible, given the capabilities of the metafile interpreter. The spacing used between character bodies is evaluated exactly; the character body, for this purpose, is an ideal character body, calculated precisely from the text aspects and the font dimensions. The position of the resulting text extent parallelogram is determined by the TEXT ALIGNMENT and the text position. Clipping is performed at least on a character by character basis.

stroke: The complete text string in the requested text font is displayed at the text position by applying all text aspects. The text string is clipped exactly at the clipping rectangle.

'Stroke' precision does not necessarily mean vector strokes; as long as the representation adheres to the rules governing stroke precision, the font may be realized in any form, for example by raster fonts.

The TEXT attributes also apply to the RESTRICTED TEXT primitive. Because determination of the text extent of a string is generally not possible in a metafile environment, the RESTRICTED TEXT element has as a parameter the size of a text restriction box. The text restriction box is a parallelogram which is derived from this parameter and the current values of the CHARACTER ORIENTATION and TEXT ALIGNMENT elements. All of the specified text string (from the RESTRICTED TEXT element and any associated APPEND TEXT elements) shall fit within the text restriction box, and the text extent of the displayed string shall not exceed the box.

If the text string as displayed with the current text attributes would exceed the box, then the values of the text attributes CHARACTER EXPANSION FACTOR, CHARACTER SPACING, TEXT FONT INDEX, TEXT PRECISION, and CHARACTER HEIGHT which are used for the display of this string are adjusted in an implementation dependent manner to achieve the required restriction. The adjustment of attributes pertains only to the restricted string, and is applied conceptually to the "realized" attribute values, i.e., those values that are actually used for the display of the string.

4.7.7 Colour Attributes

The CGM uses the RGB additive colour model used in many video devices and in colour television.

The CGM provides two mechanisms for colour selection: 'direct' and 'indexed'. In 'direct' colour selection, the colour is defined by providing values for the normalized weights of the RGB components. In 'indexed' colour selection, the colour is defined by an index into a table of direct colour values. Selection of one of these mechanisms may be done by an element in each Picture Descriptor.

For 'indexed' colour selection, the COLOUR TABLE attribute element is provided for changing the contents of the colour table. This element may appear throughout the picture body. However, the effect of changes in the colour table on any existing graphical primitive elements that use the affected indices is not addressed in this Standard.

For direct colour specification, colour values are a 3-tuple of values providing the normalized weight of the red, green, and blue components of the desired colour. In the abstract, each component of the 3-tuple is normalized to the continuous range of real numbers $[0,1]$; the normalization also has the property that any 3-tuple with 3 identical components, (x,x,x), represents equal weights of the red, green, and blue components. For any given component, one end of the range indicates that none of that component is included, and the other end indicates that the the maximum intensity of that component included in the colour, with an infinite number of component values in between. $(0,0,0)$ thus represents black, $(1,1,1)$ represents white, and (x,x,x) with x between 0 and 1 represent grays.

In a metafile, the abstract mimimum colour value of $(0,0,0)$ is represented by (min_red, min_green, min_blue) and the abstract maximum colour value of $(1,1,1)$ is represented by (max_red, max_green, max_blue). There is a Metafile Descriptor element, COLOUR VALUE EXTENT, which allows metafile generators to specify the minimum and maximum metafile colour values.

4.7.8 Filled-Area Attributes

Separate control is provided over the appearances of the interior and the edge of filled-area primitives.

The INTERIOR STYLE attribute selects one of five styles in which the interiors of filled-area elements are rendered:

hollow: No filling, but the boundary (bounding line) of the filled area is drawn using the fill colour currently selected (either via FILL BUNDLE INDEX or FILL COLOUR depending on the corresponding FILL COLOUR ASF). The boundary of a 'hollow' filled area is considered to be the representation of the interior. The boundary is distinct from the edge, and is drawn only for 'hollow' filled areas. The linetype and linewidth of the boundary are implementation dependent.

solid: Fill the interior using the fill colour currently selected (either via FILL BUNDLE INDEX or FILL COLOUR depending on the corresponding FILL COLOUR ASF).

pattern: Fill the interior using the pattern index currently selected (either via FILL BUNDLE INDEX or PATTERN INDEX, depending on the corresponding PATTERN INDEX ASF) as an index into the pattern table.

hatch: Fill the interior using the fill colour and the hatch index currently selected (either via FILL BUNDLE INDEX or individual attributes FILL COLOUR and HATCH INDEX, depending on the corresponding ASFs).

empty: No filling is done and no boundary is drawn, i.e., nothing is done to represent the interior. The only potentially visible component of an 'empty' filled area is the edge, subject to EDGE VISIBILITY and the other edge attributes.

The edge can be either visible or invisible. If visible, the individual edge attributes or the EDGE BUNDLE INDEX (according to the edge ASF values) govern the appearance.

If the edge is visible, it is drawn on top of the interior — the edge has precedence over the interior when drawn, and will always be fully visible. The boundary drawn for style 'hollow' is considered as the representation of the interior. While the edge has precedence, the boundary may be partly visible as well. Parts of edges which are clipped become invisible — clipping of edges is identical to clipping of line elements. Parts of interiors which are clipped will, in the case of style 'hollow', have a boundary drawn at the clipping boundary.

The "realized edge" is defined to be the zero-width ideal boundary line of the filled-area if the edge is invisible, and the finite-width displayed line if the edge is visible. This Standard does not mandate the alignment of the finite-width realized edge with respect to the zero-width ideal edge (i.e., whether the former is centred on the latter or aligned some other way such as inside).

The "realized interior" is defined as extending to and terminating at the realized edge. The discussion of interior in the remainder of this Standard should be considered to pertain to realized interior.

4.8 Escape Elements

ESCAPE elements describe device- or system-dependent data in the CGM. ESCAPEs may be included in the metafile at the discretion of the user, but direct effects and side effects of the use of nonstandardized elements are beyond the scope of this Standard. This Standard imposes no constraints on the functional intent or content of data passed by the ESCAPE mechanism.

4.9 External Elements

External elements communicate information not directly related to the generation of a graphical image. They may appear anywhere in the CGM.

The MESSAGE element specifies a string of characters used to communicate information to operators at CGM interpretation time. This element is intended to be used to provide special device-dependent information necessary to process a CGM. Control over the position and appearance of the character string is not provided.

The APPLICATION DATA element allows applications to store and access private data. This element is not a graphical element and its interpretation will have no effect on any picture produced by an interpreter.

For specification of non-standardized graphical effects, the ESCAPE and GENERALIZED DRAWING PRIMITIVE elements are provided. These elements may have an effect on the picture produced by an interpreter.

4.10 Conceptual State Diagram

There are a number of required sequential relationships between metafile elements, which determine whether it is syntactically correct. For example, the Metafile Descriptor (which is the first sequence of consecutive elements classified as Metafile Descriptor elements) shall occur in a metafile after the BEGIN METAFILE element and before any other elements (disregarding external and escape elements).

Conceptually, any metafile generator or interpreter may consider that the sequence of pictures and actions represented by the metafile imply changes of state in a virtual device. The valid sequences of metafile elements can therefore documented by means of a state diagram.

For the purposes of illustrating state relationships, the only significant capability of this hypothetical device is the ability to traverse the metafile data structure from beginning to end and to identify or comprehend (as opposed to interpret, render, or display) the metafile elements. The only significant structural component of the machine is a "state register". The identification by this abstract machine of various metafile elements in the sequential data structure causes the state register to assume certain values. The metafile "states" in the state diagrams (figure 12) can then be considered to be the values of the state register of this abstract device.

This presentation of metafile states and explanation in terms of abstract interpreters is solely for the purpose of illustrating and clarifying the rules of sequentiality of metafile elements. It is in no way intended to mandate the behaviour or structure of actual metafile generators and interpreters — standardization of generators and interpreters is beyond the scope of this Standard.

4.11 Registration

For certain elements, the CGM defines value ranges of parameters as being reserved for registration. The meanings of these values will be defined using the established procedures of the ISO International Registration Authority for Graphical Items. These procedures do not apply to values and value ranges defined as being reserved for implementation-dependent or private use; these values and ranges are not standardized.

Applications therefore shall not use parameter values in the reserved ranges for implementation-dependent or private use. Those metafile elements that will be affected by registration of graphical items are:

a)   LINE TYPE;

b)   MARKER TYPE;

c)   HATCH STYLE;

d)   EDGE TYPE;

e)   FONT LIST;

f)   GENERALIZED DRAWING PRIMITIVE;

g)   ESCAPE.

Registration of character sets for use with the CHARACTER SET LIST element is according to the procedures established by ISO 2375, Character Set Registration.

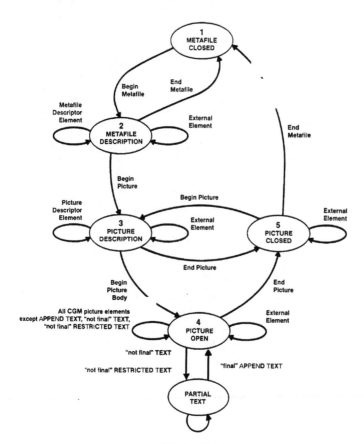

Figure 12. State diagram.

ANSI X3.122 - 1986 Part 1

5 Abstract Specification of Elements

5.1 Introduction

The metafile elements are discussed in this clause.

The Delimiter Elements (described in sub-clause 5.2) delimit significant structures within the Metafile.

The Metafile Descriptor Elements (described in sub-clause 5.3) describe the functional content, default conditions, identification, and characteristics of the CGM.

The Picture Descriptor Elements (described in sub-clause 5.4) define the extent of VDC space and declare the parameter modes of attribute elements for the entire picture.

The Control Elements (described in sub-clause 5.5) specify size and precision of coordinate space and format descriptions of the CGM elements.

The Graphical Primitive Elements (described in sub-clause 5.6) describe geometric objects in the CGM.

The Attribute Elements (described in sub-clause 5.7) describe the appearance of graphical primitive elements.

The Escape Elements (described in sub-clause 5.8) describe device- and system-dependent elements used to construct a picture.

The External Elements (described in sub-clause 5.9) communicate information not directly related to the generation of a graphical image.

The format used throughout this clause to define the Metafile element set separates functionality from coding. Each element is named, the parameters are described, data types are listed, and a description of implicit relationships is added to clarify how the element fits into the system.

Abbreviations of data type names:

Data Types		Meaning
CI	Colour Index	Non-negative integer pointer into a table of colour values.
CD	Colour Direct	Three-tuple of non-negative real values for red, green, blue colour intensities.
E	Enumerated	Set of standardized values. The set is defined by enumerating the identifiers that denote the values.
I	Integer	Number with no fractional part.
IX	Index	Integer pointer into a table of values, or integer used to select from among a set of enumerated values.
P	Point	Two VDC values representing the x and y coordinates of a point in VDC space.
R	Real	Number with integer and fractional portion, only one of which need exist.
S	String	Sequence of characters.
VDC	VDC value	Single real or integer values (as determined by VDC TYPE) in VDC space.
D	Data Record	User-defined and otherwise non-standardized record of data that accompanies elements such as APPLICATION DATA, ESCAPE, and GENERALIZED DRAWING PRIMITIVE.

Type IX parameters used as enumeration selectors in some elements have a fixed number of values with defined and standardized meanings, and have other values available for implementation-dependent definition and use. It is anticipated that the standardized values will be expanded in future versions of the CGM. To avoid possible conflict with user-defined values, the standardized and user-available values are assigned to distinct ranges of the IX parameter. Negative values of IX are allocated for private or implementation-dependent meanings, and non-negative values are reserved for (future) standardization.

Type E (enumerated) parameters also are extensible for private values, but the method of specifying private values is encoding dependent.

Combinations of simple types can also be used where n is an unspecified number (for example, nP or 2R,2I). Also, lists of types can be expressed (for example, I,E,R,E).

How these data types are represented in a given encoding of the CGM is specified in the subsequent encoding parts of this Standard.

This Standard defines the syntax and semantics of the elements that may occur in a metafile. It does not standardize the behaviour of metafile generators or interpreters. Hence it does not specify how metafile interpreters should respond to errors in the content of a metafile.

To aid designers of metafile interpreters in achieving uniformity of results, three categories of errors and degeneracies in metafile contents are identified. Annex D identifies some of these conditions and contains suggestions for reasonable responses.

The three categories of errors and degeneracies, and the way they are dealt with in the Standard, are as follows:

a) Syntactic: Errors of syntax include such conditions as the wrong amount of data for an element, or negative values in a parameter whose legal values are non-negative. While this clause does generally state valid values for element parameters, it does not attempt to enumerate exceptions. Annex D contains a general discussion, with a few examples, of such degeneracies. It does not contain an element-by-element treatment of such conditions.

b) Geometric degeneracies: Geometric degeneracies include elements that are properly specified syntactically and mathematically, but whose defining data yields a geometrically degenerate result. The principle class of geometric degeneracy is zero-length or zero-area primitives. Examples of such conditions are coincidence of all defining points in a polyline and coincidence of all edge segments in a polygon.

This Standard does not specify whether such degeneracies should produce a visible result or not. As with syntactic errors, this clause does not contain identification of specific degeneracies in the descriptions of the individual elements. In annex D, there are general recommendations for implementors who wish such degeneracies to produce visible results.

c) Mathematical singularities and ambiguities: Exceptions of this category include elements whose defining data yield a mathematically ill-specified or ambiguous result. Three colinear points in 3-point specified circular arc elements are examples in this category. This clause identifies such conditions and refers to annex D for further discussion. Annex D describes, on an element-by-element basis, such conditions and recommends mathematically sensible reactions where such exist.

5.2 Delimiter Elements

5.2.1 BEGIN METAFILE

Parameters:

identifier (S)

Description:
This is the first element of a metafile. It demarcates the beginning of the Metafile Descriptor. BEGIN METAFILE shall occur exactly once in a metafile. The identifier parameter is available for use by metafile generators and interpreters in a manner that is not further standardized.

NOTE - If more than one CGM is to be recorded on the same output medium, each shall be addressed by reference to its BEGIN METAFILE element.

This element causes a state transition in the state diagram of figure 12, into the METAFILE DESCRIPTION state.

References:
4.2

5.2.2 END METAFILE

Parameters:

None

Description:
This is the last element of a metafile.

NOTE - This element causes a state transition in the state diagram of figure 12, into the METAFILE CLOSED state.

References:
4.2

5.2.3 BEGIN PICTURE

Parameters:

identifier (S)

Description:
This is the first element of a picture. It demarcates the beginning of the Picture Descriptor. It forces all elements to return to the default values. The identifier parameter is available for use by metafile generators and interpreters in a manner that is not further standardized.

For compatibility with ISO 2022 designating and invoking controls which may occur within the string parameters of TEXT, APPEND TEXT, RESTRICTED TEXT and GENERALIZED DRAWING PRIMITIVE elements, the way that BEGIN PICTURE forces the character set to assume its default value is as follows:

BEGIN PICTURE causes the character set selected by the default value of CHARACTER SET INDEX to be designated as the current G0 set and invoked into positions 2/1 through 7/14 of the 7-bit or 8-bit code chart.

BEGIN PICTURE also designates the character set selected by the default value of ALTERNATE CHARACTER SET INDEX as the current G1 set and also as the current G2 set.

In an 8-bit environment, BEGIN PICTURE invokes that default G1 set into code chart positions 10/1 through 15/14 (or 10/0 through 15/15 if the G1 set should be a 96-character set).

Here, the terms "designate", "invoke", "G0 set", "G1 set", and "G2 set" have the meanings defined for them in ISO 2022.

NOTE - BEGIN PICTURE and END PICTURE bound the set of elements of a single picture in the CGM. Every picture in a metafile is totally independent from every other picture and always starts with a BEGIN PICTURE. This independence is enforced by returning the modal values of all elements to their default values at the start of the picture.

This element causes a state transition in the state diagram of figure 12, into the PICTURE DESCRIPTION state.

References:
4.2

5.2.4 BEGIN PICTURE BODY

Parameters:

None

Description:
This element demarcates the end of the Picture Descriptor and the beginning of the body of the picture. It thus informs the metafile interpreter of the transition from the Picture Descriptor to the graphical primitive, attribute, and control elements that define the picture.

If a new picture begins with a cleared view surface, the initial colour of the view surface is the colour specified by the BACKGROUND COLOUR element, if that element is present in the Picture Descriptor, or by the default background colour, if the BACKGROUND COLOUR element is not present in the Picture Descriptor.

Each picture defines a graphical image independent of the other pictures. As suggested in annex D, presentation of each picture on a cleared view surface is the most expected action. Because view surface clearing is not standardized, interpreters are free to compose images by overlaying pictures.

NOTE - This element causes a state transition in the state diagram of figure 12, into the PICTURE OPEN state.

References:
4.2
D.4.1

5.2.5 END PICTURE

   Parameters:

   None

   Description:
   This is the last element of a picture.

   Only external and escape elements may occur between END PICTURE and BEGIN PICTURE or between END PICTURE and END METAFILE.

   NOTE - No explicit actions are specified to occur when this element is encountered.

   This element causes a state transition in the state diagram of figure 12, into the PICTURE CLOSED state.

   References:
   4.2
   *D.*4.1

5.3 Metafile Descriptor Elements

5.3.1 METAFILE VERSION

Parameters:

version (I)

Description:
The metafile conforms to the specified version of the CGM Standard. This element shall occur in the Metafile Descriptor of every metafile.

This version of the CGM standard is version one (1). Subsequent versions of the CGM standard will use higher numbered versions.

References:
4.3.1

5.3.2 METAFILE DESCRIPTION

Parameters:

description (S)

Description:
The contents of the metafile are described in a non-standardized way by this entry.

NOTE - This element allows the CGM to be identified with descriptive text such as author, place of origin, etc.

References:
4.3.1

5.3.3 VDC TYPE

Parameters:

vdc type (one of: integer, real) (E)

Description:
The single parameter is an enumerative value that declares the data type, integer or real, of the Virtual Device Coordinates.

References:
4.3

5.3.4 INTEGER PRECISION

Parameters:

The form of the parameter depends on the specific encoding.

Description:
The precision for operands of data type integer (I) is specified for subsequent data of type I. The precision is defined as the field width measured in units applicable to the specific encoding.

References:
4.3

5.3.5 REAL PRECISION

Parameters:

The form of the parameter depends on the specific encoding.

Description:
The precision for operands of data type real (R) is specified for subsequent data of type R. The precision is defined as the field width measured in units applicable to the specific encoding. The precision may consist of parameters that define subfields of data type R.

References:
4.3

5.3.6 INDEX PRECISION

Parameters:

The form of the parameter depends on the specific encoding.

Description:
The precision for operands of data type index (IX) is specified for subsequent data of type IX. The precision is defined as the field width measured in units applicable to the specific encoding.

References:
4.3

5.3.7 COLOUR PRECISION

Parameters:

The form of the parameter depends on the specific encoding.

Description:
The precision for operands of data type colour direct (CD) is specified for subsequent data of type

CD. The precision is defined as the field width measured in units applicable to the specific encoding.

Although the form of the parameter is encoding dependent, the parameter is a single specification that applies to each or all of the three components (red, green, blue) of parameters of type CD. The precisions of the individual components are not independently and differently specifiable by this element.

References:
4.3

5.3.8 COLOUR INDEX PRECISION

Parameters:

The form of the parameter depends on the specific encoding.

Description:
The precision for operands of data type colour index (CI) is specified for subsequent data of type CI. The precision is defined as the field width measured in units applicable to the specific encoding.

References:
4.3

5.3.9 MAXIMUM COLOUR INDEX

Parameters:

maximum colour index (CI)

Description:
The parameter represents an upper bound (not necessarily the least upper bound) on colour index values that will be encountered in the metafile.

References:
4.3

5.3.10 COLOUR VALUE EXTENT

Parameters:

minimum colour value (CD)
maximum colour value (CD)

Description:
The parameters represent an extent which bounds the direct colour values that will be encountered in the metafile. It need not represent the exact extent of colour values contained in the metafile.

The 'minimum colour value' corresponds to the abstract RGB specification of (0,0,0), which means zero intensity of each of the RGB components and represents black. The 'maximum colour value' corresponds to the abstract RGB specification of (1,1,1), which means maximum intensity of each of the RGB components and represents white.

References:
4.7.7

5.3.11 METAFILE ELEMENT LIST

Parameters:

The form of the parameter is encoding dependent.

Description:
All of the elements that may be encountered in the metafile and that are not mandatory are listed. (Mandatory elements are those which shall be contained in every syntactically correct metafile.) METAFILE ELEMENT LIST shall occur in the Metafile Descriptor of every metafile.

Shorthand names are provided for use in the metafile elements list. These names may be used in conjunction with individual element names in the element list. These are:

DRAWING SET
DRAWING PLUS CONTROL SET

The elements included in each of these sets are listed in sub-clause 4.3.2.

NOTE - The information carried by this element can be used by the interpreters to determine the maximum facilities necessary for interpreting the metafile. The list represents an upper bound of functional capability. It need not be the least upper bound. Every element in the metafile shall be in the list, but the list may include elements not found in the metafile.

References:
4.3.2

5.3.12 METAFILE DEFAULTS REPLACEMENT

Parameters:

Picture Descriptor, Control, and Attribute elements.

Description:
Each element in the element list will have the same format, meaning, and parameter data types as it does when it occurs outside the METAFILE DEFAULTS REPLACEMENT element. Clause 6 gives default values for those CGM elements for which defaults make sense. Substitute or replacement values for the defaults may be defined with the METAFILE DEFAULTS REPLACE-MENT. Any subset of the elements given defaults in clause 6 may be included. Each picture within this metafile assumes that at BEGIN PICTURE the modal values of all elements are the default values whether the defaults are the clause 6 values or come from this element.

The parameters in the defaults replacement list are order dependent. When an element is encountered in the defaults replacement list, the value replaces the current default value for the element. If an element occurs more than once in the defaults replacement list, then the last value specified is the default value used by BEGIN PICTURE.

The content and format for elements in the default list are the same as the content and format for setting corresponding elements. The format of the parameter list is not further elaborated here in order to allow freedom for encodings to treat this complex element in the manner best suited to the encodings.

When a value has more than one specification mode, this standard defines its default for each mode. An element that sets a default value in the defaults replacement list shall set the value in the current specification mode. The current specification-mode when processing the list is either the default mode defined by this Standard, or the mode most recently set by an element in the list. The list may contain element that set values in more than one mode.

Elements in the list are processed sequentially. If a value is defined more than once in the list, the default that actually takes effect is the one set latest in the list.

References:
4.3.3

5.3.13 FONT LIST

Parameters:

font names (nS)

Description:
This element permits selection of named fonts via TEXT FONT INDEX. The first font defined in the font list is assigned to index 1. The second to index 2, etc.

NOTE - The strings may contain registered names or private names. Use of the former is recommended for metafile transportability, because registration ensures unique naming of fonts.

Fonts are registered in the ISO International Register of Graphical Items, which is maintained by the Registration Authority. When a font has been approved by the ISO Working Group on Computer Graphics, the font name will be assigned by the Registration Authority.

References:
4.7.6

5.3.14 CHARACTER SET LIST

Parameters:

list of:
    character set type - (one of: 94-character G-set,
                                    96-character G-set,
                                    94-character multibyte G-set,
                                    96-character multibyte G-set,
                                    complete code) (E),
    designation sequence tail (S)

Description:
The CHARACTER SET LIST element declares the character sets that can be named in subsequent CHARACTER SET INDEX and ALTERNATE CHARACTER SET INDEX elements and establishes the character set index value that is associated with each of these character sets.

The first character set declaration in the list names the character set whose character set index value is to be 1. Likewise, the second, third, fourth, etc., character set declarations name the character sets whose index values are to be 2, 3, 4, etc.

Each character set declaration has two parts: an enumerated parameter and a short string parameter. The enumerated type specifies which type of character set is being declared (that is, which type of ISO 2022 designating escape sequence is associated with that character set). The string consists of the character that forms the "tail end" of such designating escape sequences for that character set.

NOTE - There are five types of character sets: 94-character G-sets, 96-character G-sets, 94-character multibyte G-sets, 96-character multibyte G-sets, and character sets intended to be designated as "complete codes".

94-CHARACTER G-SETS. These character sets are designated by ISO 2022 escape sequences of the form <ESC> <I1> <I>(o) <F>. Here, <I1> is either 2/8, 2/9, 2/10, or 2/11; <I>(o) represents zero or more intermediate characters from column 2 of the code chart; and <F> is a final character from columns 3 through 7 of the code chart. If <F> is from column 3 of the code chart, the character set is a "private" character set. If <F> is from columns 4 through 7 of the code chart, the character set is a "standard" character set in the sense that it and its designating escape sequences are registered in the International Register Of Coded Character Sets To Be Used With Escape Sequences.

For 94-character G-sets, the character set declaration consists of 94 character, followed by a string consisting of all characters in the ISO 2022 designating escape sequence except the first two characters, <ESC> <I1>.

For example, the G-set from the U.K.'s national 7-bit character set is registered in the International Register Of Coded Character Sets To Be Used With Escape Sequences. Its designating escape sequences are as follows:

<ESC>	2/8	4/1	{to designate it as G0}
<ESC>	2/9	4/1	{to designate it as G1}
<ESC>	2/10	4/1	{to designate it as G2}
<ESC>	2/11	4/1	{to designate it as G3}

Again, the French character set (1982 version, from the 1982 version of AFNOR NF Z 62-010) is registered in the International Register Of Coded Character Sets To Be Used With Escape Sequences. Its designating escape sequences are as follows:

<ESC>	2/8	6/6	{to designate it as G0}
<ESC>	2/9	6/6	{to designate it as G1}
<ESC>	2/10	6/6	{to designate it as G2}
<ESC>	2/11	6/6	{to designate it as G3}

Therefore, a CHARACTER SET LIST element could specify that the U.K. character set is to be referred to by character set index 1, and the French character set by character set index 2, as follows:

<CHARACTER-SET-LIST: U.K., French>

'94-character G-set' 4/1
'94-character G-set' 6/6

96-CHARACTER G-SETS. These character sets are similar to 94-character G-sets, but include the code positions 2/0 and 7/15, which are excluded from 94-character G-sets. Their ISO 2022 designating escape sequences take the form <ESC> <I1> <I>(o) <F>, where the first intermediate character <I1> is either 2/13, 2/14, or 2/15. The remainder of the escape sequence is similar to the escape sequences for 94-character G-sets: zero or more intermediate characters from column 2 of the code chart and a final character from columns 3 through 7 of the code chart.

For 96-character G-sets, the character set declaration consists of '96-character G-set', followed by a string consisting of all characters in the ISO 2022 designating escape sequence except the first two characters, <ESC> <I1>.

So far, no 96-character G-sets of graphic characters have been registered in the International Register Of Coded Character Sets To Be Used With Escape Sequences. However, it is possible for interchanging parties to agree on a private 96-character G-set whose designating escape sequences would end with a character from column 3 of the code chart. For example, the following might be private escape sequences to designate such a G-set:

<ESC>	2/13	3/0	{to designate it as G1}
<ESC>	2/14	3/0	{to designate it as G2}
<ESC>	2/15	3/0	{to designate it as G3}

(96-character G-sets may not be designated as G0 sets.)

For example, the following CHARACTER SET LIST element establishes the U.K. 94-character G-set, the French 94-character G-set, and a private 96-character G-set as the character sets named by character set indexes 1, 2, and 3, respectively:

<CHARACTER-SET-LIST: U.K., French, private 96-character G-set>

'94-character G-set' 4/1
'94-character G-set' 6/6
'96-character G-set' 3/0

94-CHARACTER MULTIBYTE G-SETS. A 94-character multibyte G-set can contain 94 to the Nth power characters, each coded as a sequence of N bytes from columns 2 through 8 of the code chart, not including the bytes 2/0 and 7/15, which are excluded from 94-character G-sets. For example, a 94-character 2-byte G-set can contain 8,836 characters.

The ISO 2022 designating escape sequences for 94-character multibyte G-sets have the following forms:

<ESC>	2/4	<F>		{to designate it as G0}
<ESC>	2/4	2/9	<F>	{to designate it as G1}
<ESC>	2/4	2/10	<F>	{to designate it as G2}
<ESC>	2/4	2/11	<F>	{to designate it as G3}

For 94-character multibyte G-sets, the character set declaration consists of '94-character multibyte G-set', followed by a string consisting only of the final character in the ISO 2022 designating escape sequence.

For example, a Japanese 2-byte character set of 6802 graphic characters has been registered in the International Register Of Coded Character Sets To Be Used With Escape Sequences, and its designating escape sequences have the form shown above, with the final character <F> being 4/0. Thus, the following CHARACTER SET LIST element could be used to specify that this 2-byte Japanese character set is to be referred to by character set index 2:

<CHARACTER-SET-LIST: Japanese 2-byte character set>

'94-character multibyte G-set' 4/0

96-CHARACTER MULTIBYTE G-SETS OF GRAPHIC CHARACTERS. A 96-character multibyte G-set is similar to a 94-character multibyte G-set except that it can include the bytes 2/0 and 7/15. Thus, a 96-character 2-byte G-set could have 96 times 96 (or 9216) 2-byte character codes.

The ISO 2022 designating escape sequences for 96-character multibyte G-sets have the following forms:

<ESC>	2/4	2/13	<F>	{to designate it as G1}
<ESC>	2/4	2/14	<F>	{to designate it as G2}

<ESC>   2/4   2/15   <F>   {to designate it as G3}

It is not possible to designate a 96-character multibyte G-set as a G0 set.

The character set declaration for a 96-character multibyte G-set consists of '96-character multi-byte G-set' followed by a string consisting only of the final character <F> in the character set's ISO 2022 designating escape sequence.

So far, no 96-character multibyte G-sets have been registered in the International Register of Coded Character Sets To Be Used with Escape Sequences.

CHARACTER SETS INTENDED TO BE DESIGNATED AS COMPLETE CODES. Other character sets may not fit the ISO 2022 "G-set" structure. ISO 2022 provides an escape sequence format for invoking coding systems different from ISO 2022. The compete code escape sequences have the following form:

<ESC> 2/5 <I>o <F>

where <I>o means "zero or more characters from column 2 of the code chart", and <F> is a final character from columns 3 through 7 of the code chart. If <F> is from column 3, the coding system is a private code. If <F> is from columns 4 through 7, it is a code for which a designating and invoking escape sequence has been registered in the International Register Of Coded Character Sets To Be Used With Escape Sequences.

The character set declaration for a character set that would be invoked as a coding system different from ISO 2022 consists of 'complete code' followed by a string consisting only of those characters in the code's ISO 2022 escape sequence which come after the first two characters, <ESC> 2/5.

As well as of using a registered complete code, a private code could be used. For example, suppose the interchanging parties have agreed on a private 8-bit code to be invoked by the following escape sequence:

<ESC> 2/5 2/0 3/0

The following CHARACTER SET LIST element would declare the French character set to have character set index 1 and that 8-bit private code to have character set index 2:

<CHARACTER-SET-LIST: French, private coding system>

'96-character G-set' 6/6
'complete code' 2/0 3/0

Information regarding the designation sequence tail parameter can be found in the International Register of Coded Character Sets to be Used with Escape Sequences. This register is maintained by the Registration Authority for ISO 2375, which is the European Computer Manufacturers Association (ECMA), Rue du Rhone 114, CH-1204, Geneva, Switzerland.

References:
4.7.6

5.3.15 CHARACTER CODING ANNOUNCER

Parameters:

coding technique (E)

Description:
This element informs the metafile interpreter of the code extension capabilities assumed by the metafile generator.

'Coding technique' identifies the code extension technique and environment assumed by the generator of the metafile. These code extension capabilities apply only to the string parameters of TEXT, APPEND TEXT, RESTRICTED TEXT and possible GENERALIZED DRAWING PRIMITIVE (GDP) elements. Whether 'coding technique' applies to string parameters within the data record of a given GDP depends upon the definition of the particular GDP. The standardized values are:

BASIC 7-BIT

Character sets are switched by using CHARACTER SET INDEX, which designates a set into G0.

If ALTERNATE CHARACTER SET INDEX appears in the METAFILE ELEMENT LIST, it signals that the G1 set may be accessed using SI/SO as described in ISO 2022.

BASIC 8-BIT

Character sets are switched by using CHARACTER SET INDEX and ALTERNATE CHARACTER SET INDEX.

The G1 set may be accessed by characters from columns 10 through 15 of an 8-bit code chart. No locking or single shifts are used within the text string.

EXTENDED 7-BIT

Sets G0, G1, G2, and G3 may be invoked using the 7-bit encoding of any of the locking shifts or single shifts, in conformance with ISO 2022. CHARACTER SET INDEX selects G0 and ALTERNATE CHARACTER SET INDEX selects both G1 and G2. Designation of G2 and G3 may be done within text strings, in conformance with ISO 2022. (Designation of G0 and G1 may not be done in this fashion.)

EXTENDED 8-BIT

Sets G0, G1, G2, and G3 may be invoked using the 8-bit encoding of any of the locking shifts or single shifts, in conformance with ISO 2022. CHARACTER SET INDEX selects G0 and ALTERNATE CHARACTER SET INDEX selects both G1 and G2. Designation of G2 and G3 may be done within text strings, in conformance with ISO 2022. (Designation of G0 and G1 may not be done in this fashion.)

If text strings are coded with any other technique, this shall be announced with a private value.

NOTE - This element corresponds to the "announcer" sequences of ISO 2022.

References:
4.7.6

5.4 Picture Descriptor Elements

5.4.1 SCALING MODE

Parameters:

scaling mode (one of: abstract, metric) (E)
metric scale factor (R)

Description:
The scaling mode parameter defines the meaning of the VDC. If set to 'abstract', the VDC space is dimensionless and the picture is correctly displayed at any size: the metric scale factor parameter is ignored. If set to 'metric', the VDC space has implied measure: the metric scale factor represents the distance (in millimetres) on the in the displayed picture corresponding to one VDC unit. One VDC unit represents one millimeter multiplied by the metric scale factor. In this case the picture is correctly displayed at the indicated size only. If used, SCALING MODE shall appear in the Picture Descriptor, after BEGIN PICTURE and before BEGIN PICTURE BODY.

References:
4.4.1

5.4.2 COLOUR SELECTION MODE

Parameters:

colour selection mode (one of: indexed, direct) (E)

Description:
Two methods of colour selection are supported: by colour table entries ('indexed') or by red, green, and blue colour values ('direct').

Only one colour mode may be used within a picture. The mode may be defaulted or explicitly set with the COLOUR SELECTION MODE element. All occurrences of colour-setting elements (AUXILIARY COLOUR, LINE COLOUR, MARKER COLOUR, FILL COLOUR, EDGE COLOUR, TEXT COLOUR) as well as the colour lists of CELL ARRAY and PATTERN TABLE shall be in the current mode. If used, COLOUR SPECIFICATION MODE shall be in the Picture Descriptor, after BEGIN PICTURE and before BEGIN PICTURE BODY.

References:
4.4.2

5.4.3 LINE WIDTH SPECIFICATION MODE

Parameters:

line width specification mode (one of: absolute, scaled) (E)

Description:
Two methods of directly specifying line width are supported: absolute measure in VDC

('absolute'), or a scaling factor to be applied to the device-dependent nominal line width at metafile interpretation time.

Only one line width mode may be used within a picture. The mode may be defaulted or may be set explicitly with the LINE WIDTH SPECIFICATION MODE element. If used, LINE WIDTH SPECIFICATION MODE shall be in the Picture Descriptor, after BEGIN PICTURE and before BEGIN PICTURE BODY. All occurrences of line width elements shall have parameters in the current mode.

References:
4.4.3

5.4.4 MARKER SIZE SPECIFICATION MODE

Parameters:

marker size specification mode (one of: absolute, scaled) (E)

Description:
Two methods of directly specifying marker size are supported: absolute measure in VDC ('absolute'), or a scaling factor to be applied to the device-dependent nominal marker size at metafile interpretation time.

Only one marker size mode may be used within a picture. The mode may be defaulted or may be set explicitly with the MARKER SIZE SPECIFICATION MODE element. If used, MARKER SIZE SPECIFICATION MODE shall be in the Picture Descriptor after the BEGIN PICTURE element and before BEGIN PICTURE BODY. All occurrences of marker size elements shall have parameters in the current mode.

References:
4.4.3

5.4.5 EDGE WIDTH SPECIFICATION MODE

Parameters:

edge width specification mode (one of: absolute, scaled) (E)

Description:
Two methods of directly specifying edge width are supported: absolute measure in VDC ('absolute'), or a scaling factor to be applied to the device-dependent nominal edge width at metafile interpretation time.

Only one edge width mode may be used within a picture. The mode may be defaulted or may be set explicitly with the EDGE WIDTH SPECIFICATION MODE element. If used, EDGE WIDTH SPECIFICATION MODE shall be in the Picture Descriptor after the BEGIN PICTURE element and before BEGIN PICTURE BODY. All occurrences of edge width elements shall have parameters in the current mode.

References:
4.4.3

5.4.6 VDC EXTENT

Parameters:

first corner (P)
second corner (P)

Description:
The two corners define a rectangular extent in VDC space that is the "region of interest" for the succeeding CGM elements.

The first corner represents the lower-left corner of the picture, and the second corner represents the upper-right corner of the picture as seen by the viewer of the picture. The values of the coordinates for any dimension may be either increasing or decreasing from the first to the second corner. For example, for devices with an upper-left origin, a picture may be described in coordinates that map directly to the device but still may be displayed correctly on a device with a lower-left origin.

The VDC EXTENT thus establishes the sense and orientation of VDC space (that is, the directions of the positive x (+x) and positive y (+y) axes, and whether the +y axis is 90-degrees clockwise or 90-degrees counterclockwise from the +x axis). See 4.4.4 and figure 1.

In particular, VDC EXTENT establishes the direction of positive and negative angles as follows: positive 90-degrees is defined to be the right angle from the positive x-axis to the positive y-axis.

Note that some attributes such as text attributes (for example, the directions of the up and base component vectors of CHARACTER ORIENTATION and, therefore, the meaning of the enumerative values 'right', 'left', 'up', 'down') are intimately bound to these definitions.

NOTE - Specification of values outside VDC EXTENT in parameters of CGM elements is permitted. VDC EXTENT demarcates the region of interest within the picture; the visible portion of an image should be contained within VDC EXTENT.

References:
4.4.4
4.4.5

5.4.7 BACKGROUND COLOUR

Parameters:

colour value (CD)

Description:
The colour value defines the background colour for the image whose definition begins with the next BEGIN PICTURE BODY element.

The single parameter of BACKGROUND COLOUR is always RGB, regardless of the current value of COLOUR SELECTION MODE. If the current COLOUR SELECTION MODE is indexed, then the BACKGROUND COLOUR element defines the initial representation of colour index 0 for the picture as well as the image background colour.

References:
4.4.6

5.5 Control Elements

5.5.1 VDC INTEGER PRECISION

Parameters:

The form of the parameter depends on the specific encoding.

Description:

The indicated precision for operands of data type point (P) and operands of data type VDC value (VDC) is specified for subsequent data of type P and of type VDC when VDC type is 'integer'. The precision is defined as the field width measured in units applicable to the specific encoding. The precision may consist of parameters that defined subfields of data types P and VDC when VDC TYPE is 'real'.

NOTE - This element enables metafiles to change the form of parameters in other metafile elements in the middle of a picture so that more efficient storage of data can be used when less precision is required.

References:
4.5.1

5.5.2 VDC REAL PRECISION

Parameters:

The form of the parameter depends on the specific encoding.

Description:

The indicated precision for operands of data type point (P) and operands of data type VDC value (VDC) is specified for subsequent data of type P and of type VDC. The precision is defined as the field width measured in units applicable to the specific encoding. The precision may consist of parameters that define subfields of data type P and VDC.

NOTE - This element enables metafiles to change the form of parameters in other metafile elements in the middle of a picture so that more efficient storage of data can be used when less precision is required.

References:
4.5.1

5.5.3 AUXILIARY COLOUR

Parameters:

auxiliary colour specifier

if the colour selection mode is 'indexed',
auxiliary colour index　(CI)

if the colour selection mode is 'direct',
auxiliary colour value   (CD)

Description:
The auxiliary colour index or value is set as specified by the parameter.

The auxiliary colour is applied to drawing of primitives as described under the TRAN-SPARENCY element, when TRANSPARENCY is 'off'.

Interpretation of this element is implementation dependent. Some recommendations are provided in annex D.

References:
D.4.4

5.5.4 TRANSPARENCY

Parameters·

transparency indicator (one of: off, on) (E)

Description:
The transparency indicator is set as specified by the parameter. TRANSPARENCY controls the application of AUXILIARY COLOUR to the drawing of subsequent primitives.

When TRANSPARENCY is 'off', the following primitives are affected as described:

  a) line elements: When LINE TYPE is non-solid, the dashes and dots are drawn in the current LINE COLOUR as usual, and the spaces between are drawn in the AUXILIARY COLOUR.

  b) POLYMARKER: For devices that display markers within raster cells, pixels that are not part of the marker definition are displayed in the AUXILIARY COLOUR.

  c) text elements: for devices that display TEXT within raster cells, pixels within the character cell that are not part of the character definition are displayed in the AUXILIARY COLOUR.

  d) filled-area elements: when INTERIOR STYLE is 'hatch', pixels in the interior of the filled-area element that are not on a hatch line are displayed in the AUXILIARY COLOUR; when EDGE TYPE is non-solid, the dashes and dots are drawn in the current EDGE COLOUR as usual, and the spaces between are drawn in the AUXILIARY COLOUR.

When TRANSPARENCY is 'on', the portions of the above primitives that would be drawn in AUXILIARY COLOUR when TRANSPARENCY is 'off' are rendered transparently, i.e., nothing is drawn in that portion of the primitive when the primitive is drawn.

Interpretation of this element is implementation dependent. Some recommendations are provided in annex D.

References:
None

5.5.5 CLIP RECTANGLE

Parameters:

first corner (P)
second corner (P)

Description:
The two corner points define the clip rectangle in VDC space.

When CLIP INDICATOR is 'on', only the portions of graphics elements inside or on the boundary of the clip rectangle are drawn.

References:
4.5.2
D. 4.4

5.5.6 CLIP INDICATOR

Parameters:

clip indicator (one of: off, on) (E)

Description:
When CLIP INDICATOR is 'off', clipping of graphical primitive elements is not required.

When CLIP INDICATOR is 'on', only the portions of graphics elements inside or on the boundary of the clip rectangle are drawn.

NOTE - It is implementation and interpreter dependent whether or not clipping is done to some limit such as VDC EXTENT or display surface boundaries even when CLIP INDICATOR is 'off'. Such action is not precluded by this Standard, and may be handled by the interpreter in accord with the particular needs of the implementation and driven device(s).

References:
4.5.2

5.6 Graphical Primitive Elements

5.6.1 POLYLINE

Parameters:

point list (nP)

Description:
A line is drawn from the first point in the parameter list to the second point, from the second point to the next point, ..., and from the next-to-last point to the last point.

References:
4.6
4.6.1

5.6.2 DISJOINT POLYLINE

Parameters:

point list (nP)

Description:
A line is drawn from the starting point to the second point, from the third point to the fourth point, from the fifth point to the sixth point, ... forming a series of disjoint single line segments.

The appearance of DISJOINT POLYLINE is controlled by the line element attributes.

NOTE - This element allows significant data compression for applications wishing to perform line pattern generation or vector polygon fill prior to metafile generation in a graphics system, and for other applications such as drawing grids.

References:
4.6
4.6.1
4.7.1

5.6.3 POLYMARKER

Parameters:

point list (nP)

Description:
The marker corresponding to the currently selected marker type is drawn at each of the points in the point list. If the marker type is one of the five predefined markers, it is drawn centred at each of the points. Other implementation-dependent markers may have other alignments where desired.

If the resulting marker is completely within the clipping area, the entire marker is drawn. If the marker position is outside the clipping rectangle, nothing is displayed. If the marker position is within the clipping rectangle but any part of the marker is outside the clipping area, then the portion of the marker within the clipping rectangle is displayed and the display of the portion outside the rectangle is device or interpreter dependent.

References:
4.6
4.6.2
4.7.2

5.6.4 TEXT

Parameters:

point (P)
flag (one of: not final, final) (E)
string (S)

Description:

The character codes specified in the string are interpreted to obtain the associated symbols from the currently selected character set. Characters are displayed on the view surface as specified by the text attributes. Format effector control characters (such as CR, LF, BS, HT, VT, and FF) are permitted in a string but their interpretation is implementation-dependent. Control characters used for character set invocation and designation (SI, SO, ESC, SS2, and SS3) are permitted according to the setting of CHARACTER CODING ANNOUNCER.

The characters are dimensioned according to the CHARACTER HEIGHT and CHARACTER EXPANSION FACTOR and are oriented according to CHARACTER ORIENTATION. The direction of the character placement in the string relative to CHARACTER ORIENTATION is according to TEXT PATH.

The flag parameter is used to permit changing the following text attributes and control elements within a string which will be aligned as a single block: TEXT FONT INDEX, TEXT PRECISION, CHARACTER EXPANSION FACTOR, CHARACTER SPACING, TEXT COLOUR, CHARACTER HEIGHT, CHARACTER SET INDEX, ALTERNATE CHARACTER SET INDEX, TEXT BUNDLE INDEX, AUXILIARY COLOUR, and TRANSPARENCY.

If the flag is set to 'not final', the character codes in the string parameter are accumulated, along with the current attribute settings. In this case, only the attribute setting elements listed above are allowed between this element and the APPEND TEXT element. With the exception of the ESCAPE element, no other metafile elements of any type are allowed. ESCAPE is permitted but has no standardized effect.

If the flag is set to 'final', the string parameter constitutes the entire string to be displayed. The position of the string relative to the text point parameter is according to TEXT ALIGNMENT. Text elements with a null string parameter are legal and may be followed by the allowed text attributes and APPEND TEXT as described above.

NOTE - When the flag is 'not final' this element causes a state transition in the state diagram of figure 12, into the PARTIAL TEXT state.

References:
4.6
4.6.3
4.7.3

4.7.6

5.6.5 RESTRICTED TEXT

Parameters:

> extent: delta width, delta height (2VDC)
> point (P)
> flag (one of: not final, final) (E)
> string (S)

Description:

RESTRICTED TEXT behaves as does TEXT, with the exception that the text is constrained to be within a parallelogram determined by the 'extent' parameter, the position, and the text attributes.

The character codes specified in the string are interpreted to obtain the associated symbols from the currently selected character set. Characters are displayed on the view surface as specified by the text attributes. Format effector control characters (such as CR, LF, BS, HT, VT, and FF) are permitted in a string but their interpretation is implementation-dependent. Control characters used for character set invocation and designation (SI, SO, ESC, SS2, and SS3) are permitted according to the setting of CHARACTER CODING ANNOUNCER.

The characters are dimensioned according to the CHARACTER HEIGHT and CHARACTER EXPANSION FACTOR and are oriented according to CHARACTER ORIENTATION. The direction of the character placement in the string relative to CHARACTER ORIENTATION is according to TEXT PATH.

The first component of the 'extent' parameter is measured parallel to the base vector of CHARACTER ORIENTATION, and the second component is measured parallel to the up vector. A parallelogram, the text restriction box, is formed whose sides are parallel to the vectors, and which have lengths as in the extent parameter. The box is placed at the position point and aligned as per the current TEXT ALIGNMENT.

All text in the string is displayed within the resulting positioned box. If necessary, and only if necessary, the values of the text attributes CHARACTER HEIGHT, CHARACTER EXPANSION FACTOR, CHARACTER SPACING, TEXT PRECISION, and TEXT FONT INDEX which are used to display this string are varied to achieve the required restriction. It is only the realized values of these attributes, used to display this single string, which are varied. The method of varying the attributes is implementation dependent.

The flag parameter is used to permit changing the following text attributes and control elements within a string which will be aligned as a single block: TEXT FONT INDEX, TEXT PRECISION, CHARACTER EXPANSION FACTOR, CHARACTER SPACING, TEXT COLOUR, CHARACTER HEIGHT, CHARACTER SET INDEX, ALTERNATE CHARACTER SET INDEX, TEXT BUNDLE INDEX, AUXILIARY COLOUR, and TRANSPARENCY.

If the flag is set to 'not final', the character codes in the string parameter are accumulated, along with the current attribute settings. In this case, only the attribute setting elements listed above are allowed between this element and the APPEND TEXT element. With the exception of the ESCAPE element, no other metafile elements of any type are allowed. ESCAPE is permitted but has no standardized effect.

If the flag is set to 'final', the string parameter constitutes the entire string to be displayed. It is this complete string to which the text restriction box applies. The position of the string relative to the text point parameter is according to TEXT ALIGNMENT. Text elements with a null string parameter are legal and may be followed by the allowed text attributes and APPEND

TEXT as described above.

NOTE - TEXT PRECISION is included in the attributes which may be changed to achieve the text restriction because TEXT PRECISION controls the relationship between currently set values of text attributes and the values actually used for display of a string (the "realized" values). The realization of the text restriction required by the RESTRICTED TEXT element may mandate another mapping from requested to realized attribute values than would be allowable under the current TEXT PRECISION. Hence the requirements of the current TEXT PRECISION may have to be ignored to achieve proper display of the RESTRICTED TEXT element.

When the flag is 'not final' this element causes a state transition in the state diagram of figure 12, into the PARTIAL TEXT state.

References:
4.6
4.6.3
4.7.3
4.7.6
D.4.5

5.6.6 APPEND TEXT

Parameters:

flag (one of: not final, final) (E)
string (S)

Description:

The character codes specified in the string are appended to the string defined by preceding nonfinal TEXT, RESTRICTED TEXT, and APPEND TEXT elements. The codes are interpreted to obtain the associated symbols from the current character set.. Characters are displayed on the view surface as specified by the text attributes. Format effector control characters (such as CR, LF, BS, HT, VT, and FF) are permitted in a string but their interpretation is implementation-dependent. Control characters used for character set invocation and designation (SI, SO, ESC, SS2, and SS3) are permitted according to the setting of CHARACTER CODING ANNOUNCER.

The characters are dimensioned according to the CHARACTER HEIGHT and CHARACTER EXPANSION FACTOR and are oriented according to CHARACTER ORIENTATION. The direction of the character placement in the string relative to CHARACTER ORIENTATION is according to TEXT PATH.

The flag parameter is used to permit changing the following text attributes and control elements within a string which will be aligned as a single block: TEXT FONT INDEX, TEXT PRECISION, CHARACTER EXPANSION FACTOR, CHARACTER SPACING, TEXT COLOUR, CHARACTER HEIGHT, CHARACTER SET INDEX, ALTERNATE CHARACTER SET INDEX, TEXT BUNDLE INDEX, AUXILIARY COLOUR, and TRANSPARENCY.

If the flag is set to 'not final', the character codes in the string parameter are accumulated, along with the current attribute settings. In this case, only the attribute setting elements listed above are allowed between this element and the APPEND TEXT element. With the exception of the external element ESCAPE, no other metafile elements of any type are allowed.

If the flag is set to 'final', the accumulated string parameter constitutes the entire string to be displayed. APPEND TEXT elements with a null string parameter are legal and may be followed by the allowed text attributes and further APPEND TEXT elements as described above.

NOTE - When the flag is 'final' this element causes a state transition in the state diagram of figure 12, into the PICTURE OPEN state.

References:
4.6
4.6.3
4.7.3
4.7.6

5.6.7 POLYGON

Parameters:

point list (nP)

Description:
A boundary of a polygonal region is defined by connecting each vertex to its successor in the ordered point list with straight edges and connecting the last vertex to the first. The polygonal region may be nonsimple. For example, edges are allowed to cross. In this way, subareas can be created. The interior of the polygon is as defined in 4.6.4.4.

A non-degenerate polygon (one with three or more vertices, not all of which are colinear) is displayed with interior as defined by the FILL BUNDLE INDEX, ASPECT SOURCE FLAGS, interior style attributes, AUXILIARY COLOUR and TRANSPARENCY. The appearance of the edge is controlled by the edge attributes and by AUXILIARY COLOUR and TRANSPARENCY.

References:
4.6
4.6.4
4.7.4
4.7.8

5.6.8 POLYGON SET

Parameters:

List of:

point (P)
edge out flag (one of: invisible, visible, close invisible, close visible) (E)

Description:
A set of closed polygons is drawn (according to the edge visibility flags and the current edge attributes) and filled (according to the current filled-area attributes).

The list of points and flags is processed sequentially. The first point starts the first polygon of the set; the point that starts each polygon is recorded as the "current closure point". Each point in the list is connected either to its successor or to the current closure point (but not both) by a straight edge.

The edge out flag parameter associated with each point in the list defines how the edge coming from that point is generated. The enumerations of the flag mean:

invisible: the edge from point n to point n+1 defines a fill boundary, and is not drawn.

visible: the edge from point n to point n+1 defines a fill boundary, and is drawn.

close invisible: the edge from point n to the current closure point defines a polygon boundary, but is not drawn. The next point in the list (if any) will define the first point of another polygon; it will not be connected by any edge to any point of the polygon being closed.

close visible: as close invisible, but the closing edge added is drawn.

If the edge out flag of the last point in the list is 'visible', it is treated as 'close visible'; if the flag is 'invisible', it is treated as 'close invisible'.

The interior of the polygon set (see 4.6.4.4) is filled according to the current filled-area attributes. The set of polygons is filled according to the parity (odd or even) algorithm described under the POLYGON element, with the exception that the transition from a vertex marked 'close visible' or 'close invisible' to the next point in the point list does not constitute a boundary to the fill algorithm.

The individual polygons of the set are not filled individually. The polygons in the set may be disjoint (as in the 'dot' and the body of the letter 'i'), may create 'holes' (as in a torus shape), or may overlap.

The visible edges are drawn using the current edge attributes. An edge will be drawn only if it was generated with either a 'visible' or 'close visible' flag and EDGE VISIBILITY is set to ON. EDGE VISIBILITY thus acts as an override on the visibility of the edges specified in the polygon set, in that it can turn off edges which were specified as 'visible', but cannot turn on edges which were specified as 'invisible'.

See figure 13 for an example of POLYGON SET.

NOTE - The ability to intermix visible and invisible edges is provided to accomodate clipping of polygons before they are placed in the CGM; a clipped edge is typically invisible.

References:
4.6.4
4.7.8
D.4.5

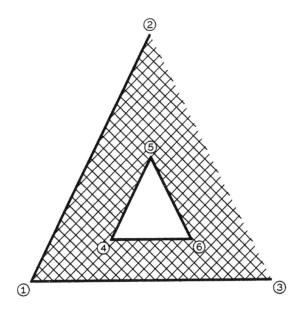

Figure 13. Example for POLYGON SET.

This figure would be generated by the following sequence, where (n) is a symbolic notation for the coordinate corresponding to the circled point n on the picture."

    (1) visible
    (2) invisible
    (3) close visible
    (4) visible
    (5) visible
    (6) close visible

5.6.9 CELL ARRAY

Parameters:

3 corner points P, Q, and R (3P)

nx,ny (2I)

local colour precision (form depends upon specific encoding)

cell colour specifiers

    if the colour selection mode is 'indexed',
    colour indexes (nx*nyCI)

    if the colour selection mode is 'direct',
    colour values (nx*nyCD)

Description:

In the general case, P,Q,R can delimit an arbitrary parallelogram. P and Q delimit the end points of a diagonal of the parallelogram, and R defines a third corner.

In the simplest case, the three corner points, P,Q,R, define a rectangular area in VDC space. This area is subdivided into nx*ny contiguous rectangles as follows. The edge from P to R is subdivided into nx equal intervals, and the edge from R to Q is subdivided into ny equal intervals. The grid implied consists of nx*ny identical cells. The colour list consists of nx*ny colour specifications, conceptually an array of dimensions nx and ny representing respectively the column and row dimensions. Array element (1,1) is mapped to the cell at corner P, and array element (nx,1) is mapped to the cell at corner R. Array element (nx,ny) is mapped to the cell at corner Q. Hence, the colour elements are mapped within rows running from P to R, and with the rows incrementing in order from R to Q.

The 'local colour precision' parameter declares the precision of the 'cell colour specifiers'. The precision is for either indexed or direct colour, according to the COLOUR SELECTION MODE of the picture. As with the COLOUR INDEX PRECISION and COLOUR PRECISION elements, the form of the parameter is encoding dependent. If the picture uses indexed colour selection, then the form of the parameter is the same as that of COLOUR INDEX PRECISION. If the picture uses direct colour selection, then the form of the parameter is the same as that of COLOUR PRECISION.

Legal values of the 'local colour precision' include the legal values of COLOUR (INDEX) PRECISION. In addition, each encoding defines a special value, the 'default colour precision indicator', as an indicator that the colour specifiers of the element are to be encoded in the COLOUR (INDEX) PRECISION of the metafile, i.e., to indicate that the 'local colour precision' defaults to COLOUR (INDEX) PRECISION.

Recommendations for the interpretation of cell array for devices that cannot display a cell array are given in annex D.

NOTE - Figure 14 illustrates a cell array where the order of mapping the cells to a display surface corresponds to the common left-to-right, top-to-bottom pixel scan order of many devices.

References:
4.6
4.6.5
D.4.5

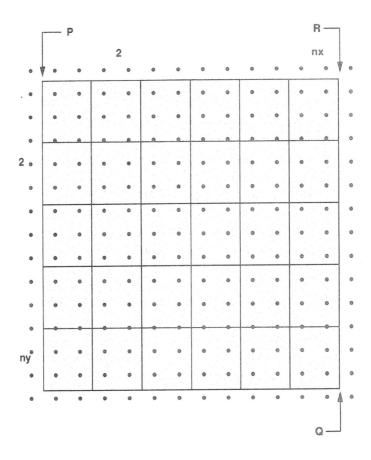

Figure 14. nx-by-ny rectangle mapped onto display surface. Lines indicate cell array locations. Dots indicate pixels.

5.6.10 GENERALIZED DRAWING PRIMITIVE (GDP)

Parameters:

identifier (I)
point list (nP)
data record (D)

Description:

A Generalized Drawing Primitive (GDP) of the type specified by the identifier is generated on the basis of the given points and the data record.

Non-negative values of the identifier are reserved for registration and future standardization, and negative values are available for private use.

The appearance of the GDP is determined by zero or more of the attribute sets of the standardized graphical primitive elements, depending on the particular GDP. The parameters of the GDP are interpreted and utilized in an interpreter dependent manner.

NOTE - GDP provides convenient access to non-standardized graphical primitives that a device may support. GDP is similar to ESCAPE in this sense, but GDP provides a mechanism for handling of coordinate data whereas ESCAPE does not. GDP is thus preferable for generating graphical output, and ESCAPE is designed for such applications as non-standardized control functions.

When registration of GDPs occurs there may be registered GDPs which correspond with some of the standardized metafile graphical primitive elements, e.g., CIRCLE.

GDP identifiers are registered in the ISO International Register of Graphical Items, which is maintained by the Registration Authority. When a GDP identifier has been approved by the ISO Working Group on Computer Graphics, the GDP identifier value will be assigned by the Registration Authority.

References:
4.6

5.6.11 RECTANGLE

Parameters:

two points (2P)

Description:

The two points specified represent diagonally opposite corners of a rectangle oriented parallel to the VDC axes. The rectangle so defined is displayed with interior (see 4.6.4.4) as defined by the FILL BUNDLE INDEX, ASPECT SOURCE FLAGS, AUXILIARY COLOUR, TRANSPARENCY, and interior style attributes. The appearance of the edge is controlled by the edge attributes, and by AUXILIARY COLOUR and TRANSPARENCY.

References:
4.6
4.6.4
4.7.4
4.7.8

5.6.12 CIRCLE

Parameters:

centrepoint (P)
radius (VDC)

Description:
A circle of the specified radius at the specified centrepoint is displayed with interior (see 4.6.4.4) as defined by the FILL BUNDLE INDEX, ASPECT SOURCE FLAGS, AUXILIARY COLOUR, TRANSPARENCY, and interior style attributes. The appearance of the edge is controlled by the edge attributes, and by AUXILIARY COLOUR and TRANSPARENCY.

Valid values of 'radius' are non-negative VDC.

References:
4.6
4.6.4
4.7.4
4.7.8

5.6.13 CIRCULAR ARC 3 POINT

Parameters:

starting point, intermediate point, ending point (3P)

Description:
A circular arc is displayed from the starting point, through the specified intermediate point, to the specified ending point.

A non-degenerate specification is one in which the three specified coordinates are non-colinear.

If the three specified coordinates are colinear the specification is mathematically degenerate, and the interpretation of this element is implementation dependent. See annex D for further discussion.

References:
4.6
4.6.1
4.6.6
4.7.1
D. 4.5

5.6.14 CIRCULAR ARC 3 POINT CLOSE

Parameters:

starting point, intermediate point, ending point (3P)
close type (one of: pie, chord) (E)

Description:

A filled circular arc is displayed from the specified starting point through the specified intermediate point, to the specified ending point. The close types are illustrated in figure 15.

If close type is 'chord', the segment defined by the arc and the chord from the starting point to the ending point is displayed with interior (see 4.6.4.4) as defined by the FILL BUNDLE INDEX, ASPECT SOURCE FLAGS, AUXILIARY COLOUR, TRANSPARENCY, and interior style attributes. The appearance of the edge is controlled by the edge attributes, and by AUXILIARY COLOUR and TRANSPARENCY.

If close type is 'pie', the pie sector defined by the computed arc centre, the specified starting point, and the ending point is displayed with interior as defined by the FILL BUNDLE INDEX, ASPECT SOURCE FLAGS, AUXILIARY COLOUR, TRANSPARENCY, and interior style attributes. The appearance of the edge is controlled by the edge attributes, and by AUXILIARY COLOUR and TRANSPARENCY.

A non-degenerate specification is one in which the three specified coordinates are non-colinear.

If the three specified coordinates are colinear the specification is mathematically degenerate and ambiguous, and the interpretation of this element is implementation dependent. See annex D for further discussion.

References:
4.6
4.6.4
4.6.6
4.7.4
4.7.8
D. 4.5

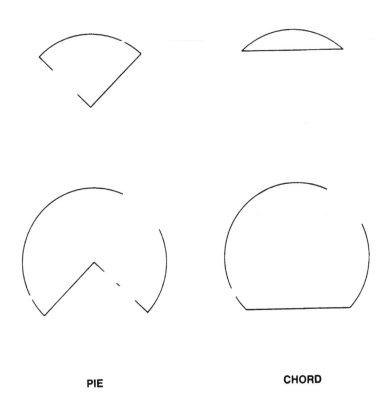

PIE   **CHORD**

Figure 15. CIRCULAR ARC 3 POINT CLOSE specifications with 'pie' and 'chord'.

5.6.15 CIRCULAR ARC CENTRE

Parameters:

    centrepoint (P)
    DX_start,DY_start,DX_end,DY_end (4VDC)
    radius (VDC)

Description:

A circular arc is drawn which is defined as follows:

DX_start and DY_start define a start vector, and DX_end and DY_end define an end vector. The tails of these vectors are placed on the centrepoint. A start ray and end ray are derived from the start and end vectors. The start and end rays are the semi-infinite lines from the centrepoint in the directions of the start and end vectors respectively.

The specified radius and centrepoint define a circle. The arc is drawn in the positive angular direction (as defined by VDC EXTENT) from the intersection of the circle and the start ray (as obtained by measuring a distance 'radius' along the start ray from the centrepoint) to the intersection of the circle and the end ray.

The arc is displayed with current line element attributes.

Valid values of the vector components are those which produce vectors of non-zero length.

Valid values of 'radius' are non-negative VDC.

If the start ray and end ray are coincident, it is ambiguous whether the defined arc subtends 0 degrees or 360 degrees of central angle. See annex D for further discussion.

References:
    4.6
    4.6.1
    4.6.6
    4.7.1
    D.4.5

5.6.16 CIRCULAR ARC CENTRE CLOSE

Parameters:

    centrepoint (P)
    DX_start,DY_start,DX_end,DY_end (4VDC)
    radius (VDC)
    close type (one of: pie, chord) (E)

Description:

A circular arc is drawn and filled which is defined as follows:

DX_start and DY_start define a start vector, and DX_end and DY_end define an end vector. The tails of these vectors are placed on the centrepoint. A start ray and end ray are derived from the start and end vectors. The start and end rays are the semi-infinite lines from the centrepoint in the directions of the start and end vectors respectively.

The specified radius and centrepoint define a circle. The arc is drawn in the positive angular direction (as defined by VDC EXTENT) from the intersection of the circle and the start ray (as obtained by measuring a distance 'radius' along the start ray from the centrepoint) to the intersection of the circle and the end ray.

If the CLOSE TYPE is 'chord', the circular segment defined by the arc and the chord from the starting point to the ending point of the arc is displayed.

If the CLOSE TYPE is 'pie', the circular sector defined by the arc and the specified centrepoint is displayed.

The primitive is displayed with interior (see 4.6.4.4) as defined by the FILL BUNDLE INDEX, ASPECT SOURCE FLAGS, and interior style attributes. The appearance of the edge is controlled by the edge attributes, and by AUXILIARY COLOUR and TRANSPARENCY.

Valid values of the vector components are those which produce vectors of non-zero length.

Valid values of 'radius' are non-negative VDC.

If the start ray and end ray are coincident, it is ambiguous whether the defined arc subtends 0 degrees or 360 degrees of central angle. See annex D for further discussion.

References:
    4.6
    4.6.4
    4.6.6
    4.7.4
    4.7.8
    *D.*4.5

5.6.17 ELLIPSE

Parameters:

centrepoint (P)
first CDP endpoint (P)
second CDP endpoint (P)

Description:
The centrepoint specifies the center of an ellipse. The CDP endpoints include one endpoint from each conjugate diameter; together with the centrepoint they define the two conjugate diameters of the ellipse.

The ellipse so specified is displayed with interior (see 4.6.4.4) as defined by the FILL BUNDLE INDEX, ASPECT SOURCE FLAGS, AUXILIARY COLOUR, TRANSPARENCY, and interior style attributes. The appearance of the edge is controlled by the edge attributes, and by AUXILIARY COLOUR and TRANSPARENCY.

Valid values of the three specifying points of the ellipse are those which yield three distinct points. The specified ellipse is non-degenerate if and only if the three points are non-colinear.

References:
    4.6
    4.6.4
    4.6.7
    4.7.4
    4.7.8
    *D.*4.5

5.6.18 ELLIPTICAL ARC

Parameters:

centrepoint (P)
first CDP endpoint (P)
second CDP endpoint (P)
DX_start,DY_start,DX_end,DY_end (4VDC)

Description:
An elliptical arc is drawn which is defined as follows:

The centrepoint specifies the center of an ellipse. The CDP endpoints include one endpoint from each conjugate diameter; together with the centrepoint they define the two conjugate diameters of the ellipse.

DX_start and DY_start define a start vector, and DX_end and DY_end define an end vector. The tails of these vectors are placed on the centrepoint. A start ray and end ray are derived from the start and end vectors. The start and end rays are the semi-infinite lines from the centrepoint in the directions of the start and end vectors respectively.

The define arc begins at the intersection of the ellipse and the start ray and follows the ellipse to the intersection of the ellipse and the end ray in the direction defined as follows. A "conjugate radius" is defined to be half of a conjugate diameter. Letting the centrepoint be labelled M, the first CDP endpoint P1, and the second CDP endpoint P2, then the line segments M-P1 and M-P2 define two conjugate radii, referred to in what follows as the first conjugate radius and the second conjugate radius respectively. The conjugate radii meet at M and define two angles: the sum of the two angles is 360 degrees, one angle is less than 180 degrees and the other is greater than 180 degrees. The drawing direction of the elliptical arc is the direction from the first conjugate radius to the second conjugate radius through the smaller of these two angles.

The elliptical arc is displayed with the current line attributes.

Valid values of the three specifying points of the ellipse are those which yield three distinct points. The specified ellipse is non-degenerate if and only if the three points are non-colinear.

Valid values of the vector components are those which produce vectors of non-zero length. If the start ray and end ray are coincident, it is ambiguous whether the defined arc is null (zero arc length) or the entire ellipse. See annex D for further discussion.

References:
4.6
4.6.1
4.6.7
4.7.1
D.4.5

5.6.19 ELLIPTICAL ARC CLOSE

Parameters:

centrepoint (P)
first CDP endpoint (P)
second CDP endpoint (P)
DX_start,DY_start,DX_end,DY_end (4VDC)
close type (one of: pie, chord) (E)

Description:
An elliptical arc is drawn and filled which is defined as follows:

The centrepoint specifies the center of an ellipse. The CDP endpoints include one endpoint from each conjugate diameter; together with the centrepoint they define the two conjugate diameters of the ellipse.

DX_start and DY_start define a start vector, and DX_end and DY_end define an end vector. The tails of these vectors are placed on the centrepoint. A start ray and end ray are derived from the start and end vectors. The start and end rays are the semi-infinite lines from the centrepoint in the directions of the start and end vectors respectively.

The define arc begins at the intersection of the ellipse and the start ray (the "starting point") and follows the ellipse to the intersection of the ellipse and the end ray (the "ending point") in the direction defined as follows. A "conjugate radius" is defined to be half of a conjugate diame-. ter. Letting the centrepoint be labelled M, the first CDP endpoint P1, and the second CDP endpoint P2, then the line segments M-P1 and M-P2 define two conjugate radii, referred to in what follows as the first conjugate radius and the second conjugate radius respectively. The conjugate radii meet at M and define two angles: the sum of the two angles is 360 degrees, one angle is less than 180 degrees and the other is greater than 180 degrees. The drawing direction of the elliptical arc is the direction from the first conjugate radius to the second conjugate radius through the smaller of these two angles.

If the close type is 'chord', the segment defined by the elliptical arc and the chord from the starting point to the ending point is displayed with interior (see 4.6.4.4) as defined by the FILL BUNDLE INDEX, FILL ASF, and interior style attributes. The appearance of the edge is controlled by edge attributes, and by AUXILIARY COLOUR and TRANSPARENCY.

If the close type is 'pie', the elliptical pie sector defined by the elliptical arc centrepoint, the starting point, and the ending point is displayed with interior as defined by the FILL BUNDLE INDEX, FILL ASF, and the interior style attributes. The appearance of the edge is controlled by the edge attributes, and by AUXILIARY COLOUR and TRANSPARENCY. The elliptical arc is displayed with the current line attributes. An elliptical arc is drawn which is defined as follows:

Valid values of the three specifying points of the ellipse are those which yield three distinct points. The specified ellipse is non-degenerate if and only if the three points are non-colinear.

Valid values of the vector components are those which produce vectors of non-zero length. If the start ray and end ray are coincident, it is ambiguous whether the defined arc is null (zero arc length) or the entire ellipse. See annex D for further discussion.

References:
4.6
4.6.4
4.6.7
4.7.4
4.7.8
D.4.5

5.7 Attribute Elements

5.7.1 LINE BUNDLE INDEX

Parameters:

line bundle index (IX)

Description:
The line bundle index is set to the value specified by the parameter. When subsequent line elements occur, the values for LINE TYPE, LINE WIDTH, and LINE COLOUR are taken from the corresponding components of the indexed bundle if the ASFs for those attributes are set to 'bundled'. See 4.6 for a list of line elements.

If the ASF for a given attribute is 'individual', this element does not affect the value used for that attribute until the ASF returns to 'bundled'.

Legal values are positive integers.

References:
4.6.1
4.7.1
D. 4.6

5.7.2 LINE TYPE

Parameters:

line type indicator (IX)

Description:
The line type indicator is set to the value specified by the parameter.

When the LINE TYPE ASF is 'individual', subsequent line elements are displayed with this line type. See 4.6 for a list of line elements.

When the LINE TYPE ASF is 'bundled', this element does not affect the display of subsequent line elements until the ASF returns to 'individual'.

The following line types are assigned:

1: solid
2: dash
3: dot
4: dash-dot
5: dash-dot-dot

Values above 5 are reserved for registration and future standardization, and negative values are available for implementation-dependent use.

NOTE - Ideally, line type ideally is maintained continuously between adjacent spans of a single POLYLINE element (see annex D for further discussion.) This consideration does not apply to other line elements, as they do not have interior defining vertices. This Standard does not specify continuity between separate, but graphically connected, line elements; nor does it specify continuity across sections of a single line element that may have been clipped away.

Whether or not line type is maintained continuously across the segments of DISJOINT POLY-LINE is not addressed by this Standard.

Line type indicator values are registered in the ISO International Register of Graphical Items, which is maintained by the Registration Authority. When a line type indicator has been approved by the ISO Working Group on Computer Graphics, the line type indicator value will be assigned by the Registration Authority.

References:
4.6.1
4.7.1
D.4.6

5.7.3 LINE WIDTH

Parameters:

line width specifier

if line width specification mode is 'absolute',
absolute line width   (VDC)

if line width specification mode is 'scaled',
line width scale factor   (R)

Description:
The absolute line width or line width scale factor is set as specified by the parameter.

When the LINE WIDTH ASF is 'individual', subsequent line elements are displayed according to the size specification of this element. See 4.6 for a list of line elements.

When the LINE WIDTH ASF is 'bundled', this element does not affect the display of subsequent line elements until the ASF returns to 'individual'.

Valid values of 'line width specifier' are non-negative VDC if LINEWIDTH SPECIFICATION MODE is 'absolute', and non-negative reals if LINEWIDTH SPECIFICATION MODE is 'scaled'.

NOTE - The line width is measured perpendicular to the defining line (that is, it is independent of the orientation of the defining line). A wide line is aligned with its ideal zero-width defining line such that the distance between the defining line and either edge is half the line width.

The appearance of lines at endpoints and interior vertices or corners (that is, whether they are mitred, rounded, etc.) is not addressed by this Standard.

References:
4.4.3
4.6.1
4.7.1
4.7.5
D.4.6

5.7.4 LINE COLOUR

Parameters:

line colour specifier

if the colour selection mode is 'indexed',
line colour index   (CI)

if the colour selection mode is 'direct',
line colour value   (CD)

Description:

The line colour index or line colour value is set as specified by the parameter(s).

When the LINE COLOUR ASF is 'individual', subsequent line elements are drawn in this line colour. See 4.6 for a list of line elements.

When the LINE COLOUR ASF is 'bundled', this element does not affect the interpretation of subsequent line elements until the ASF returns to 'individual'.

References:

4.6.1
4.7.1
4.7.7
D. 3.2

5.7.5 MARKER BUNDLE INDEX

Parameters:

marker bundle index (IX)

Description:

The marker bundle index is set to the value specified by the parameter. When subsequent marker elements occur, the values for MARKER TYPE, MARKER SIZE, and MARKER COLOUR are taken from the corresponding components of the indexed bundle if the ASFs for those attributes are 'bundled'.

If the ASF for a given attribute is 'individual', this element does not affect the value used for that attribute until its ASF returns to 'bundled'.

Legal values of MARKER BUNDLE INDEX are positive integers.

References:

4.6.2
4.7.2
D. 4.6

5.7.6 MARKER TYPE

Parameters:

marker type (IX)

Description:
The marker type is set to the value specified by the parameter.

When the MARKER TYPE ASF is 'individual', subsequent marker elements are displayed with this marker type.

When the MARKER TYPE ASF is 'bundled', this element does not affect the display of subsequent marker elements until the ASF returns to 'individual'.

The following marker types are assigned:

1: dot (.)
2: plus (+)
3: asterisk (*)
4: circle (o)
5: cross (x)

The marker type 'dot' is intended always to be displayed as the smallest visible point on the display surface at metafile interpretation time. It is thus intended to behave as a polypoint element.

Values above 5 are reserved for registration and future standardization, and negative values are available for implementation-dependent use.

NOTE - Marker type values are registered in the ISO International Register of Graphical Items, which is maintained by the Registration Authority. When a marker type has been approved by the ISO Working Group on Computer Graphics, the marker type value will be assigned by the Registration Authority.

References:
4.6.2
4.7.2
D. 4.6

5.7.7 MARKER SIZE

Parameters:

marker size specifier

if marker size specification mode is 'absolute',
absolute marker size (VDC)

if marker size specification mode is 'scaled',
marker size scale factor (R)

Description:
The absolute marker size or marker size scale factor is set as specified by the parameter. If absolute, the specified size is the maximum extent of the marker.

When the MARKER SIZE ASF is 'individual', subsequent marker elements are displayed according to the size specification of this element.

When the MARKER SIZE ASF is 'bundled', this element does not affect the display of subsequent marker elements until the ASF returns to 'individual'.

Valid values of 'marker size specifier' are non-negative VDC if MARKER SIZE SPECIFICATION MODE is 'absolute', and non-negative reals if MARKER SIZE SPECIFICATION MODE is 'scaled'.

References:
4.4.3
4.6.2
4.7.2
4.7.5
D. 4.6

5.7.8 MARKER COLOUR

Parameters:

marker colour specifier

if the colour selection mode is 'indexed',
marker colour index   (CI)

if the colour selection mode is 'direct',
marker colour value   (CD)

Description:
The marker colour index or marker colour value is set as specified by the parameter(s).

When the MARKER COLOUR ASF is 'individual', subsequent marker elements are displayed with this marker colour.

When the MARKER COLOUR ASF is 'bundled', this element does not affect the display of subsequent marker elements until the ASF returns to 'individual'.

References:
4.6.2
4.7.2
4.7.7
D. 3.2

5.7.9 TEXT BUNDLE INDEX

Parameters:

text bundle index (IX)

Description:
The text bundle index is set to the value specified by the parameter. When subsequent text elements occur, the values for TEXT FONT INDEX, TEXT PRECISION, CHARACTER EXPANSION FACTOR, CHARACTER SPACING, and TEXT COLOUR are taken from the corresponding components of the indexed bundle if the ASFs for those attributes are set to 'bundled'. See 4.6 for a list of text elements.

If the ASF for a given attribute is 'individual', this element does not affect the value used for that attribute until the ASF returns to 'bundled'.

Legal values of the text bundle index parameter are positive integers.

References:
4.6.3
4.7.3
4.7.6
D. 4.6

5.7.10 TEXT FONT INDEX

Parameters:

font index (IX)

Description:
The font index is set to the value specified by the parameter. The font index is used to select a font from the font list table defined in the Metafile Descriptor (or the default font list, if none was specified).

When the TEXT FONT INDEX ASF is 'individual', subsequent text elements are displayed with this font index. See 4.6 for a list of text elements.

When the TEXT FONT INDEX ASF is 'bundled', this element does not affect the display of subsequent text elements until the ASF returns to 'individual'.

Legal values of the font index parameter are positive integers.

NOTE - Metafile generators should ensure that the selected character set and text font are compatible.

Annex D gives recommendations for interpreters to follow in the case that the currently selected character set cannot be rendered in the specified text font.

References:
4.6.3
4.7.3
4.7.6

5.7.11 TEXT PRECISION

Parameters:

text precision (one of: string, character, stroke) (E)

Description:
The text precision is set to the value specified by the parameter.

When the TEXT PRECISION ASF is 'individual', subsequent text elements are displayed with this text precision. See 4.6 for a list of text elements.

When the TEXT PRECISION ASF is 'bundled', this element does not affect the display of subsequent text elements until the ASF returns to 'individual'.

The accuracy of execution of TEXT attributes can be controlled by one of three values.

If 'string' precision is specified, only the text position of subsequent text strings need be guaranteed, and the manner in which the string is clipped is implementation dependent.

If 'character' precision is specified, the metafile interpreter guarantees that the starting position of each character satisfy the relevant text attributes, thus guaranteeing orientation and placement of the string; however, skew, orientation, and size of each character are not guaranteed. All characters of the string which lie completely inside or outside the clipping region are clipped as appropriate, but the effect of clipping on a character whose character box is intersected by the clipping boundary is implementation dependent.

If 'stroke' precision is specified, the metafile interpreter guarantees that the placement, skew, orientation, and size of all characters satisfy all standardized text attributes. Characters are clipped to the geometric accuracy of the device.

References:
    4.6.3
    4.7.3
    4.7.6
    D. 4.6

5.7.12 CHARACTER EXPANSION FACTOR

Parameters:

character expansion factor (R)

Description:
The character expansion factor is set to the value specified by the parameter.

When the CHARACTER EXPANSION FACTOR ASF is 'individual', subsequent text elements are displayed with this character expansion factor. See 4.6 for a list of text elements.

When the CHARACTER EXPANSION FACTOR ASF is 'bundled', this element does not affect the display of subsequent text elements until the ASF returns to 'individual'.

The character expansion factor specifies the deviation of the width-to-height ratio of the character from the ratio indicated by the font designer.

Legal values of the character expansion factor are non-negative reals.

NOTE - The character expansion factor is a scalar. The resulting character width is the product of the CHARACTER HEIGHT multiplied by the width/height ratio (a characteristic of each font and a quantity that can vary on a character-by-character basis) for the character multiplied by the CHARACTER EXPANSION FACTOR. The character width so derived is further scaled by multiplying it by the ratio of the length of the character base vector to the length of the character up vector.

References:
    4.6.3
    4.7.3
    4.7.6
    D. 4.6

5.7.13 CHARACTER SPACING

Parameters:

character spacing (R)

Description:
The character spacing is set to the value specified by the parameter.

When the CHARACTER SPACING ASF is 'individual', subsequent text elements are displayed with this character spacing. See 4.6 for a list of text elements.

When the CHARACTER SPACING ASF is 'bundled', this element does not affect the display of subsequent text elements until the ASF returns to 'individual'.

The parameter represents the desired space to be added between character bodies of a text string, which is in addition to any intercharacter spacing provided by the font within the character's body. It is specified as a fraction of the current CHARACTER HEIGHT attribute. The space is added along the text path. A negative value implies that characters may overlap.

When TEXT PATH is right or left, the character spacing is scaled by the ratio of the length of the character base vector to the length of the character up vector.

References:
4.6.3
4.7.3
4.7.6
D.4.6

5.7.14 TEXT COLOUR

Parameters:

text colour specifier

if the colour selection mode is 'indexed',
text colour index   (CI)

if the colour selection mode is 'direct',
text colour value   (CD)

Description:
The text colour index or text colour value is set as specified by the parameter(s).

When the TEXT COLOUR ASF is 'individual', subsequent text elements are displayed with this text colour. See 4.6 for a list of text elements.

When the TEXT COLOUR ASF is 'bundled', this element does not affect the display of subsequent text elements until the ASF returns to 'individual'.

The text colour index is a pointer into the colour table.

References:
4.6.3
4.7.3
4.7.6

4.7.7
D.3.2

5.7.15 CHARACTER HEIGHT

Parameters:

character height (VDC)

Description:
The character height is set to the value specified by the parameter. Subsequent text elements are displayed with this character height. See 4.6 for a list of text elements.

The parameter represents the desired height of the character body, from baseline to capline, in VDC units; it is a positive number. It is measured along the character up vector. If the character orientation vectors are not orthogonal, this will not be the perpendicular distance between baseline and capline.

Valid values of 'character height' are non-negative VDC.

References:
4.6.3
4.7.6
D.4.6

5.7.16 CHARACTER ORIENTATION

Parameters:

x character up component   (VDC)
y character up component   (VDC)
x character base component (VDC)
y character base component (VDC)

Description:
The two vectors define the orientation and skew of the character body in subsequent text elements. See 4.6 for a list of text elements. For purposes of alignment and path, 'up' is in the direction of the character up vector and 'right' is in the direction of the character base vector. The ratio of the length of the base vector to the length of the up vector is used as a scaling factor for the CHARACTER EXPANSION FACTOR and CHARACTER SPACING elements.

Valid values of the vectors include any which have non-zero length, and are not collinear.

NOTE - The way in which software above the metafile generator and/or the metafile generator may use this element is as follows. A vector whose length is the character height (baseline-to-capline) and whose direction is the desired character up vector is created. A second vector is also created with the same length, whose direction is negative 90-degrees from the up vector. This pair of vectors may be transformed before being given to the metafile generator as the parameters to CHARACTER ORIENTATION. If the resultant vectors are not orthogonal, the text extent rectangle becomes a parallelogram, and the characters are skewed. If the vectors have different lengths, the aspect ratio derived from the font design and the character expansion attribute will be altered. If the positive angle from the up vector to the base vector is less than 180-degrees, the following effects occur: characters are mirror imaged; and the "intuitive" notions of right and left (as applied to TEXT PATH and TEXT ALIGNMENT) are reversed, as described in 4.7.6.

References:
4.6.3
4.7.6
D.4.6

5.7.17 TEXT PATH

Parameters:

text path (one of: right, left, up, down) (E)

Description:
The text path is set to the value specified by the parameter. Subsequent text elements are displayed with this text path. See 4.6 for a list of text elements.

This function sets the value of the text path attribute, specifying the writing direction of a text string relative to the character up vector and character base vector. 'Right' means in the direction of the character base vector. 'Left' means 180 degrees from the character base vector. 'Up' means in the direction of the character up vector. 'Down' means 180 degrees from the character up vector.

References:
4.6.3
4.7.6
D.4.6

5.7.18 TEXT ALIGNMENT

Parameters:

horizontal alignment (one of: normal horizontal, left, centre, right, continuous horizontal) (E)

vertical alignment (one of: normal vertical, top, cap, half, base, bottom, continuous vertical) (E)

continuous horizontal alignment (R)

continuous vertical alignment (R)

Description:
The text alignment is set to the value specified by the parameters. Subsequent text strings are displayed with this text alignment.

The horizontal alignment type parameter is an enumerated data type with the possible values shown above. If its value is 'continuous horizontal', the continuous horizontal alignment parameter (which is a fraction of the side of the text extent rectangle perpendicular to character up vector) becomes significant.

The vertical alignment type parameter is an enumerated data type with the possible values shown above. If the value is 'continuous vertical', the continuous vertical alignment parameter (which is a fraction of the side of the text extent rectangle parallel to character up vector) becomes significant.

The "normal" parameters are dependent on the text path at the time of the elaboration of the text elements. See 4.6 for a list of text elements.

PATH	NORMAL HORIZONTAL	NORMAL VERTICAL
RIGHT	LEFT	BASELINE
LEFT	RIGHT	BASELINE
UP	CENTRE	BASELINE
DOWN	CENTRE	TOP

The continuous horizontal and vertical parameters may exceed the range of 0.0 to 1.0 in order to align a string with a coordinate outside its text extent rectangle.

References:
4.6.3
4.7.6
D. 4.6

5.7.19 CHARACTER SET INDEX

Parameters:

character set index (IX)

Description:
The specified character set from the table specified in the CHARACTER SET LIST Metafile Descriptor element becomes the currently designated G0 set. Since BEGIN PICTURE invokes the G0 set into positions 2/1 through 7/14 of the 7-bit or 8-bit code chart, the character set designated by CHARACTER SET INDEX is used to display the text in the text elements. See 4.5 for a list of text elements. The character set is used for the subsequent mapping of character codes to character symbols.

Legal values of character set index parameter are positive integers.

NOTE - One use of this element is to switch among character sets for different languages.

References:
4.6.3
4.7.6
D. 4.6

5.7.20 ALTERNATE CHARACTER SET INDEX

Parameters:

alternate character set index (IX)

Description:
The specified character set from the table specified in the CHARACTER SET LIST Metafile Descriptor element becomes the currently designated G1 set and also the currently designated G2 set. Since BEGIN PICTURE invokes the G1 set into positions 10/1 through 15/14 (or 10/0 through 15/15 if the G1 set should be a 96-character set), the character set designated by

ALTERNATE CHARACTER SET INDEX is used to display 8-bit bytes whose most significant bit is set when those bytes occur within the string parameters of the text elements. This character set is used for the subsequent mapping of character codes to character symbols.

Legal values of the alternate character set index parameter are positive integers.

If the appropriate CHARACTER CODING ANNOUNCER is selected the SO and SI controls and ISO 2022 escape sequences may be embedded within the string parameters of text elements. If they are, the characters occurring after SO (SHIFT OUT) and before the SI (SHIFT IN) are displayed using the G1 character set: the same character set which is designated by ALTERNATE CHARACTER SET INDEX.

References:
  4.6.3
  4.7.6

5.7.21 FILL BUNDLE INDEX

Parameters:

  fill bundle index (IX)

Description:
  The fill bundle index is set to the value specified by the parameter. When subsequent filled-area elements occur, values for INTERIOR STYLE, FILL COLOUR, PATTERN INDEX, and HATCH INDEX are taken from the corresponding components of the indexed bundle, if the ASFs for these attributes are set to 'bundled'. See 4.6 for a list of filled-area elements. If the ASF for a given attribute is 'individual', this element does not affect the value used for that attribute until its ASF returns to 'bundled'.

  Legal values of FILL BUNDLE INDEX are positive integers.

References:
  4.6.4
  4.7.4
  4.7.8
  D. 4.6

5.7.22 INTERIOR STYLE

Parameters:

  interior style (one of: hollow, solid, pattern, hatch, empty) (E)

Description:
  The interior style of the filled-area elements is set to the value specified by the parameter. See 4.6 for a list of filled-area elements.

  If other non-standardized values of interior style are used, they shall be given private values.

  When the INTERIOR STYLE ASF is 'bundled', this element does not affect the display of filled-area elements until the ASF returns to 'individual'.

  The interior fill style is used to determine in what style the area is to be filled. (See 4.7.8 for discussion of interior styles, and of extent of the interior and relationship of the interior to the edge.)

References:
4.6.4
4.7.4
4.7.8
D. 4.6

5.7.23 FILL COLOUR

Parameters:

fill colour specifier

if the colour selection mode is 'indexed',
fill colour index   (CI)

if the colour selection mode is 'direct',
fill colour value   (CD)

Description:
The fill colour index or fill colour value is set as specified by the parameter(s).

When the FILL COLOUR ASF is 'individual', subsequent filled-area elements are filled with this colour. See 4.6 for a list of filled-area elements.

When the FILL COLOUR ASF is 'bundled', this element does not affect the display of these subsequent filled-area elements until the ASF returns to 'individual'.

The fill colour index is a pointer into the colour table. The fill colour attribute is significant only if INTERIOR STYLE is 'hollow', 'solid', or 'hatch'.

References:
4.6.4
4.7.4
4.7.7
4.7.8
D. 4.6

5.7.24 HATCH INDEX

Parameters:

hatch index   (IX)

Description:
The hatch index is set to the value specified by the parameter.

The following hatch indices are assigned:

1: horizontal equally spaced parallel lines
2: vertical equally spaced parallel lines
3: positive slope equally spaced parallel lines
4: negative slope equally spaced parallel lines

5: horizontal/vertical crosshatch
6: positive slope/negative slope crosshatch

The ideal angle for the positive slope hatch patterns is +45 degrees, and the ideal angle for the negative slope hatch patterns is +135 degrees. (See annex D for further discussion.)

When the HATCH INDEX ASF is 'individual' and the interior style is 'hatch', subsequent filled-area elements are displayed using this hatch index. See 4.6 for a list of filled-area elements.

When the HATCH INDEX ASF is 'bundled', this element does not affect the display of subsequent filled-area elements until the ASF returns to 'individual'.

The fill colour attribute determines the colour of the hatch lines.

Values above 6 are reserved for registration and future standardization, and negative values are available for implementation-dependent use.

NOTE - Hatch index values are registered in the ISO International Register of Graphical Items, which is maintained by the Registration Authority. When a hatch index has been approved by the ISO Working Group on Computer Graphics, the hatch index value will be assigned by the Registration Authority.

References:
4.6.4
4.7.4
4.7.8
$D.4.6$

5.7.25 PATTERN INDEX

Parameters:

pattern index (IX)

Description:
The pattern index is set to the value specified by the parameter.

When the PATTERN INDEX ASF is 'individual' and the interior style is 'pattern', subsequent filled-area elements are displayed using this pattern index. See 4.6 for a list of filled-area elements.

When the PATTERN INDEX ASF is 'bundled', this element does not affect the display of subsequent filled-area elements until the ASF returns to 'individual'.

The pattern index is a pointer into the pattern tables.

Legal values of PATTERN INDEX are positive integers.

References:
4.6.4
4.7.4
4.7.8

5.7.26 EDGE BUNDLE INDEX

Parameters:

edge bundle index (IX)

Description:
The edge bundle index is set to the value specified by the parameter. For subsequent filled-area elements, values for EDGE TYPE, EDGE WIDTH, and EDGE COLOUR are taken from the corresponding components of the indexed bundle, if the ASFs for these attributes are set to 'bundled'. See 4.6 for a list of filled-area elements. If the ASF for a given attribute is 'individual', this element does not affect the value used for that attribute until its ASF returns to 'bundled'.

Legal values of EDGE BUNDLE INDEX are positive integers.

References:
4.6.4
4.7.4
4.7.8
D. 4.6

5.7.27 EDGE TYPE

Parameters:

edge type indicator (IX)

Description:
The edge type indicator is set to the value specified by the parameter.

When the EDGE TYPE ASF is 'individual' and EDGE VISIBILITY is 'on' the edges of filled-area elements are displayed with this edge line type. See 4.6 for a list of filled-area elements.

When the EDGE TYPE ASF is 'bundled', this element does not affect the display of subsequent filled-area elements until the ASF returns to 'individual'.

Edge type indicator has the same correspondence between type (for example, 4) and representation (dash-dot) as line type indicator. The following edge types are assigned:

    1: solid
    2: dash
    3: dot
    4: dash-dot
    5: dash-dot-dot

Non-negative values of the index are reserved for standardized edge types, and negative values are available for implementation dependent use.

NOTE - Ideally, edge type is maintained continuously between adjacent spans of a single filled-area element. Continuity across edge sections that may have been clipped away or that have been declared invisible (in the case of POLYGON SET) is not addressed by this Standard.

Edge type values are registered in the ISO International Register of Graphical Items, which is maintained by the Registration Authority. When a edge type has been approved by the ISO Working Group on Computer Graphics, the edge type value will be assigned by the Registration Authority.

References:
4.6.4
4.7.4
4.7.8

D. 4.6

5.7.28 EDGE WIDTH

Parameters:

edge width specifier, either

if edge width specification mode is 'absolute',
absolute edge width   (VDC)

if edge width specification mode is 'scaled',
edge width scale factor   (R)

Description:
The absolute edge width or edge width scale factor is set as specified by the parameter.

When the EDGE WIDTH ASF is 'individual' and EDGE VISIBILITY is 'on', the edge of filled-area elements are displayed with this width. See 4.6 for a list of filled-area elements.

When the EDGE WIDTH ASF is 'bundled', this element does not affect the display of subsequent filled-area elements until the ASF returns to 'individual'.

Valid values of 'edge width specifier' are non-negative VDC if EDGE WIDTH SPECIFICATION MODE is 'absolute', and non-negative reals if EDGE WIDTH SPECIFICATION MODE is 'scaled'.

NOTE - When a edge line is displayed, the edge width is measured perpendicular to the defining line (that is, it is independent of the orientation of the defining line). No particular alignment of the finite-width displayed edge with the zero-width defining line (i.e., whether centre alignment or some other) is mandated by this Standard. See annex D for guidelines about alternatives.

References:
4.4.3
4.6.4
4.7.4
4.7.5
4.7.8
D. 4.6

5.7.29 EDGE COLOUR

Parameters:

edge colour specifier

if the colour selection mode is 'indexed',
edge colour index   (CI)

if the colour selection mode is 'direct',
edge colour value   (CD)

Description:

The edge colour index or edge colour value is set as specified by the parameter(s).

When the EDGE COLOUR ASF is 'individual' and EDGE VISIBILITY is 'on', the edge of filled-area elements are displayed with this colour. See 4.6 for a list of filled-area elements.

When the EDGE COLOUR ASF is 'bundled', this element does not affect the interpretation of subsequent filled-area elements until the ASF returns to 'individual'.

References:
4.6.4
4.7.4
4.7.7
4.7.8
D.3.2

5.7.30 EDGE VISIBILITY

Parameters:

edge visibility (one of: off,on) (E)

Description:
EDGE VISIBILITY specifies whether the edge of a filled-area element is displayed. This is independent of the display of the boundary, which is rendered when INTERIOR STYLE is 'hollow'. See 4.7.8 for the distinction between the edge and the boundary of a filled-area element. See 4.6 for a list of filled-area elements.

The edge is never displayed if the current value is 'off'. If the current value is 'on' it is displayed for all primitives except POLYGON SET. For polygon set, individual edges are displayed if and only if the current value of EDGE VISIBILITY is on and the edge flag indicates a visible edge.

References:
4.6.4
4.7.8

5.7.31 FILL REFERENCE POINT

Parameters:

reference point (P)

Description:
The fill reference point is set to the value specified by the parameter.

When the currently selected interior style is 'pattern', this value is used in conjunction with pattern size for displaying filled-area primitives.

When the currently selected interior style is 'hatch', the fill reference point provides a common origin for the hatch patterns in all subsequent hatched filled areas.

The position of the start of the pattern or hatch is defined by the fill reference point. Pattern is mapped onto the filled area by conceptually replicating the pattern definition in directions parallel to the sides of the pattern box until the interior of the complete filled area is covered.

The common origin for hatched filled areas means that separate filled areas that have the same hatch index and that abut have a visually continuous hatch pattern across all of the filled areas.

References:
4.6.4
4.7.8

5.7.32 PATTERN TABLE

Parameters:

pattern table index (IX)
nx,ny (2I)
local colour precision (form depends upon specific encoding)
pattern colour specifiers

if the colour selection mode is 'indexed',
colour index array   (nx*nyCI)

if the colour selection mode is 'direct',
colour value array   (nx*nyCD)

Description:
The representation of the specified pattern table index is defined. The representation consists of an nx-by-ny array of colours. When INTERIOR STYLE is 'pattern', the pattern is mapped onto the interior of a filled-area element as described under the PATTERN SIZE element.

Legal values of the pattern table index parameter are positive integers.

The 'local colour precision' parameter declares the precision of the 'pattern colour specifiers'. The precision is for either indexed or direct colour, according to the COLOUR SELECTION MODE of the picture. As with the COLOUR INDEX PRECISION and COLOUR PRECISION elements, the form of the parameter is encoding dependent. If the picture uses indexed colour selection, then the form of the parameter is the same as that of COLOUR INDEX PRECISION. If the picture uses direct colour selection, then the form of the parameter is the same as that of COLOUR PRECISION.

Legal values of the 'local colour precision' include the legal values of COLOUR (INDEX) PRECISION. In addition, each encoding defines a special value, the 'default colour precision indicator', as an indicator that the colour specifiers of the element are to be encoded in the COLOUR (INDEX) PRECISION of the metafile, i.e., to indicate that the 'local colour precision' defaults to COLOUR (INDEX) PRECISION.

This element may appear throughout the picture body. However, the effect of changes in the pattern table on any existing graphical primitive elements that use the affected indices is not addressed in this Standard.

References:
4.6.4
4.7.7
4.7.8

5.7.33 PATTERN SIZE

Parameters:

> pattern height vector, x component (VDC)
> pattern height vector, y component (VDC)
> pattern width vector, x component (VDC)
> pattern width vector, y component (VDC)

Description:
The pattern size is set to the values specified by the parameters.

When INTERIOR STYLE is set to 'pattern', subsequent filled-area elements are displayed using this pattern size. See 4.6 for a list of filled-area elements.

Pattern size is comprised of two vectors, a height vector and a width vector. In the general case the pattern size vectors and the FILL REFERENCE POINT define a parallelogram located in VDC, the pattern box. This pattern box is divided into cells, nx in the width vector direction and ny in the height vector direction, where nx and ny are the colour array dimensions of the pattern table entry selected by the current pattern index.

The array of colours of the current pattern is mapped onto the array of cells as follows. The colour array element (1,ny) is mapped to the pattern box cell which is located at the FILL REFERENCE POINT. Colour array elements with increasing first dimension are associated with successive cells in the direction of the width vector, and colour array elements with decreasing second dimension are associated with successive cells in the direction of the height vector. In this way, each of the nx*ny colour array elements is associated with one of the nx*ny cells of the pattern box.

Conceptually, the pattern box so defined is replicated in directions parallel to the vectors of the PATTERN SIZE element until the interior of a filled-area element to which the pattern is to be applied is completely covered. The coincidence of this imposed pattern and the interior to which it is to be applied defines the interior style for the filled-area element being displayed.

References:
4.6.4
4.7.8
D. 4.6

5.7.34 COLOUR TABLE

Parameters:

> starting index (CI)
> colour list (nCD)

Description:
The colour list elements are loaded, in the order specified, into the consecutive locations in the colour table beginning at the starting index. Only the specified colour table entries are changed. The effect of changes in the colour table on any existing graphical primitive elements that use the affected indexes is not standardized.

Legal values of the colour index are non-negative integers.

References:

4.7.7
D. 3.2

5.7.35 ASPECT SOURCE FLAGS

Parameters:

list of: pairs of
ASF type, ASF value (one of: individual, bundled) n[E,E]

Description
The designated Aspect Source Flags (ASFs) are set to the values indicated by the parameter.
The following ASF types are assigned:

> line type ASF
> line width ASF
> line colour ASF
> marker type ASF
> marker size ASF
> marker colour ASF
> text font index ASF
> text precision ASF
> character expansion factor ASF
> character spacing ASF
> text colour ASF
> interior style ASF
> fill colour ASF
> hatch index ASF
> pattern index ASF
> edge type ASF
> edge width ASF
> edge colour ASF

The Aspect Source Flags determine the attribute values that will be bound to a primitive. If the
ASF for a particular aspect of a primitive is set to 'individual', the value used is the value of the
corresponding individually specified attribute of the primitive. If the ASF is set to 'bundled', the
value used is the value of the corresponding aspect of the bundle pointed to by the current bundle index for the primitive.

ASFs are modally bound to primitives just as are other primitive attributes. Thus changing the
value of an ASF within a picture will have no retroactive effect on any previous graphical primitive element.

References:
4.7
D. 4.6

5.8 Escape Elements

5.8.1 ESCAPE

Parameters:

function identifier (I)
data record (D)

Description:
ESCAPE provides access to device capabilities not specified by this Standard. The function identifier parameter specifies the particular escape function. Non-negative values are reserved for registration and future standardization, and negative values are available for implementation dependent use.

NOTE - This element has been deliberately underspecified. Software making use of the ESCAPE element is less portable.

ESCAPE is designed for access to non-standardized control features of graphics devices, as opposed to non-standardized geometric primitives. The GENERALIZED DRAWING PRIMITIVE element is designed for specification of non-standardized primitives.

Function identifiers are registered in the ISO International Register of Graphical Items, which is maintained by the Registration Authority. When a function identifier has been approved by the ISO Working Group on Computer Graphics, the function identifier value will be assigned by the Registration Authority.

References:
4.8

5.9 External Elements

5.9.1 MESSAGE

Parameters:

action required flag (one of: no action, action) (E)
text (S)

Description:
The MESSAGE element specifies a string of characters used to communicate information to operators at Metafile interpretation time through a path separate from normal graphical output.

If the action required flag parameter is 'action', the metafile interpreter may need to pause to wait for an operator response. Because the message and an associated pause may be directed at a particular device, only the interpreter may determine if a pause is appropriate. Character set selection for the text parameter is independent of any character set selection specified by this standard.

References:
4.9

5.9.2 APPLICATION DATA

Parameters:

identifier (I)
data record (D)

Description:
This element supplements the information in the metafile in an application-dependent way. It has no effect on the picture generated by interpreting the metafile, or on the states of the metafile generator or interpreter.

The content of the identifier and data record parameters is not standardized.

NOTE - The contents of the data record may include such information as history data associated with pictures, description of algorithms used, etc.

References:
4.9

6 Metafile Defaults

This clause contains the Metafile default values that are used for those default values not explicitly set in the METAFILE DEFAULTS REPLACEMENTS element. The default values of some elements are dependent upon the values of other elements (for example, default CHARACTER HEIGHT is dependent upon VDC EXTENT). In these cases, the default of the dependent element is tied to the default of the other element, whether the latter is as defined in the table below or is defined with the METAFILE DEFAULTS REPLACEMENT elements. The value of the dependent element does not, however, change when the value of the element upon which it depends is changed explicitly by a metafile element. Rather, the value of the dependent element remains unchanged, in its default state, until explicitly changed by the occurrence of the element. See 5.3.11 for further discussion of element defaults and the action of METAFILE DEFAULTS REPLACEMENT

VDC TYPE:	integer
INTEGER PRECISION:	encoding dependent
REAL PRECISION:	encoding dependent
INDEX PRECISION:	encoding dependent
COLOUR PRECISION:	encoding dependent
COLOUR INDEX PRECISION:	encoding dependent
MAXIMUM COLOUR INDEX:	63
COLOUR VALUE EXTENT:	encoding dependent
METAFILE ELEMENT LIST:	n/a
METAFILE DEFAULTS REPLACEMENT:	n/a
FONT LIST:	for index 1, any font that can represent the nationality-independent subset of ISO 646 which is the default for CHARACTER SET LIST described below
CHARACTER SET LIST:	for index 1, any character set which includes the nationality-independent subset of ISO 646 in the positions specified in ISO 646
CHARACTER CODING ANNOUNCER:	basic 7-bit
SCALING MODE:	abstract; metric scale factor n/a
COLOUR SELECTION MODE:	indexed
LINE WIDTH SPECIFICATION MODE:	scaled
MARKER SIZE SPECIFICATION MODE:	scaled
EDGE WIDTH SPECIFICATION MODE:	scaled
VDC EXTENT:	if VDC TYPE is integer, lower left (0,0), upper right (32767,32767); if VDC TYPE is real, lower left (0.,0.), upper right (1.0,1.0)
BACKGROUND COLOUR:	device-dependent background colour
VDC INTEGER PRECISION:	encoding dependent
VDC REAL PRECISION:	encoding dependent
AUXILIARY COLOUR:	if COLOUR SELECTION MODE is 'indexed', 1; if COLOUR SELECTION MODE is 'direct', device-dependent foreground colour

Metafile Defaults

TRANSPARENCY:	on
CLIP RECTANGLE:	VDC EXTENT
CLIP INDICATOR:	on
LINE BUNDLE INDEX:	1
LINE TYPE:	1 (solid)
LINE WIDTH:	if LINE WIDTH SPECIFICATION MODE is 'absolute', 1/1000 of the longest side of the rectangle defined by default VDC EXTENT; if LINE WIDTH SPECIFICATION MODE is 'scaled', 1.0
LINE COLOUR:	if COLOUR SELECTION MODE is 'indexed', 1; if COLOUR SELECTION MODE is 'direct', device-dependent foreground colour
MARKER BUNDLE INDEX:	1
MARKER TYPE:	3 (asterisk)
MARKER SIZE:	if MARKER SIZE SPECIFICATION MODE is 'absolute', 1/100 of the longest side of the rectangle defined by default VDC EXTENT; if MARKER SIZE SPECIFICATION MODE is 'scaled', 1.0
MARKER COLOUR:	if COLOUR SELECTION MODE is 'indexed', 1; if COLOUR SELECTION MODE is 'direct', device-dependent foreground colour
TEXT BUNDLE INDEX:	1
TEXT FONT INDEX:	1
TEXT PRECISION:	string
CHARACTER EXPANSION FACTOR:	1.0
CHARACTER SPACING:	0.0
TEXT COLOUR:	if COLOUR SELECTION MODE is 'indexed', 1; if COLOUR SELECTION MODE is 'direct', device-dependent foreground colour
CHARACTER HEIGHT:	1/100 of the length of the longest side of the rectangle defined by default VDC extent
CHARACTER ORIENTATION:	0,1,1,0
TEXT PATH:	right
TEXT ALIGNMENT:	normal horizontal, normal vertical
CHARACTER SET INDEX:	1
ALTERNATE CHARACTER SET INDEX:	1
FILL BUNDLE INDEX:	1
INTERIOR STYLE:	hollow
FILL COLOUR:	if COLOUR SELECTION MODE is 'indexed', 1; if COLOUR SELECTION MODE is 'direct', device-dependent foreground colour
HATCH INDEX:	1

X:

INDEX: 1

                          1 (solid)

                          if EDGE WIDTH SPECIFICATION MODE is 'absolute',
                          1/1000 of the longest side of the rectangle defined by
                          default VDC EXTENT; if EDGE WIDTH SPECIFICA-
                          TION MODE is 'scaled', 1.0

                          if COLOUR SELECTION MODE is 'indexed', 1; if
                          COLOUR SELECTION MODE is 'direct', device-dependent
                          foreground colour

Y: off

E POINT: lower-left corner point of default VDC extent

E: 1;
                          nx=ny=1;
                          local colour precision is the (encoding-dependent) 'default
                          colour precision indicator';
                          if COLOUR SELECTION MODE is 'indexed', the default
                          colour specification is 1; if COLOUR SELECTION MODE
                          is 'direct', the default colour specification is device-
                          dependent foreground colour

                          0,dy,dx,0, where dy and dx are respectively the height and
                          width of the default VDC Extent

                          device-dependent background colour for index = 0;
                          device-dependent foreground colours for indexes greater
                          than 0

E FLAGS: all individual

7 Conformance

7.1 Forms of Conformance

This standard specifies functionality and encodings of Computer Graphics Metafiles; it does not specify operations or required capabilities of metafile generators or metafile readers. Guidelines are provided in annex D, however, for those striving for uniformity of results.

A metafile may conform to this Standard in one of two ways. Full conformance occurs when the metafile conforms to one of the encodings specified in this Standard. Functional conformance occurs when the content of the metafile corresponds exactly to the Functional Specification given in part 1 of this Standard, but a private encoding is used. These rules are expanded in the following sub-clauses.

7.2 Functional Conformance of Metafiles

A metafile is said to be functionally conforming to this Standard if the following conditions are met.

a) All graphical elements contained therein match the functionality of the corresponding elements of this Standard.

b) The sequence of elements in the metafile conforms to the relationships specified in this Standard, producing the structure specified in this Standard. For example, the metafile must begin with BEGIN METAFILE and end with END METAFILE, include exactly one metafile descriptor at the beginning which contains at least all the required elements, and so forth, as specified in this Standard.

c) No elements appear in the metafile other than those specified in the Standard, unless required for the encoding technique. All non-standardized elements are encoded using the ESCAPE element or the external elements APPLICATION DATA and MESSAGE.

7.3 Full Conformance of Metafiles

A metafile is said to be fully conforming to this Standard if the following conditions are met.

a) The metafile is functionally conforming, as specified above.

b) The metafile is encoded in conformance with one of the standardized encodings specified in this Standard.

7.4 Conformance of Other Encodings

A functionally conforming metafile may use a private encoding. While it is beyond the scope of this Standard to standardize rules for private encodings, annex B suggests minimum criteria that private encodings should meet.

Part 1

ANNEXES

A Formal Grammar of the Functional Specification

NOTE - This annex is not part of the Standard; it is included for information purposes only.

A.1 Introduction

This grammar is a formal definition of a standard CGM syntax. The encoding-independent and the encoding-dependent productions are separated, and there are subsections showing the syntax of each of the standardized encoding schemes. Details on the encoding of terminal symbols can be found in parts of this Standard that deal with the particular encoding schemes.

A.2 Notation Used

\<symbol\>	- nonterminal
\<SYMBOL\>	- terminal
\<symbol\>*	- 0 or more occurrences
\<symbol\>+	- 1 or more occurrences
\<symbol\>o	- optional (0 or 1 occurrences)
\<symbol\>(n)	- exactly n occurrences, n=2,3,...
\<symbol-1\> ::= \<symbol-2\>	- symbol-1 has the syntax of symbol-2
\<symbol-1\> ¦ \<symbol-2\>	- symbol-1 or alternatively symbol-2
\<symbol: meaning\>	- symbol with the stated meaning
{comment}	- explanation of a symbol or a production

A.3 Detailed Grammar

A.3.1 Metafile Structure

\<metafile\>	::= \<BEGIN METAFILE\> \<metafile identifier\> \<metafile descriptor\> \<metafile contents\> \<END METAFILE\>
\<metafile identifier\>	::= \<string\>
\<metafile contents\>	::= \<extra element\>* ¦ \<picture\> ¦ \<extra element\>*
\<extra element\>	::= \<external element\> ¦ \<escape element\>
\<picture\>	::= \<BEGIN PICTURE\> \<picture identifier\> \<picture descriptor element\>* \<BEGIN PICTURE BODY\>

```
                                <picture element>*
                                <END PICTURE>

<picture identifier>       ::=  <string>

<picture element>          ::=  <control element>
                             ¦  <graphical element>
                             ¦  <primitive attribute element>
                             ¦  <escape element>
                             ¦  <external element>
```

A.3.2 Metafile Descriptor Elements

```
<metafile descriptor>      ::=  <identification>
                                <characteristics>

<identification>           ::=  <METAFILE VERSION>
                                    <integer>
                                <metafile description>o

<metafile description>     ::=  <METAFILE DESCRIPTION>
                                    <string>

<characteristics>          ::=  <element list>
                                <optional descr elmt>*

<element list>             ::=  <METAFILE ELEMENT LIST>
                                    <element name>*

<optional descr elmt>      ::=  <VDC TYPE>
                                    <vdc type>
                             ¦  <MAXIMUM COLOUR INDEX>
                                    <colour index>
                             ¦  <COLOUR VALUE EXTENT>
                                    <red green blue>(2)
                             ¦  <METAFILE DEFAULTS REPLACEMENT>
                                    <element default>+
                             ¦  <FONT LIST>
                                    <font name>+
                             ¦  <CHARACTER SET LIST>
                                    <character set definition>+
                             ¦  <CHARACTER CODING ANNOUNCER>
                                    <coding technique enumerated>
                             ¦  <scalar precision>
                             ¦  <escape element>
                             ¦  <external element>

<vdc type>                 ::=  <INTEGER>
                             ¦  <REAL>

<element default>          ::=  <control element>
                             ¦  <picture descriptor element>
                             ¦  <attribute element>
                             ¦  <escape element>
```

\<font name\>	::=	\<string\>

\<character set definition\>	::=	\<char set enumerated\> \<designation sequence\>

\<index\>	::=	\<standard index value\> ¦ \<private index value\>

\<standard index value\>	::=	\<non-negative integer\>
\<non-negative integer\>	::=	\<integer\> {greater or equal to 0}
\<positive integer\>	::=	\<integer\> {greater than 0}
\<private index value\>	::=	\<negative integer\>
\<negative integer\>	::=	\<integer\> {less than 0}
\<positive index\>	::=	\<positive integer\>

\<char set enumerated\>	::=	\<94 CHAR\> ¦ \<96 CHAR\> ¦ \<MULTI-BYTE 94 CHAR\> ¦ \<MULTI-BYTE 96 CHAR\> ¦ \<COMPLETE CODE\>

\<coding technique enumerated\>	::=	\<BASIC 7-BIT\> ¦ \<BASIC 8-BIT\> ¦ \<EXTENDED 7-BIT\> ¦ \<EXTENDED 8-BIT\>

\<designation sequence\>	::=	\<string\>

\<scalar precision\>	::=	\<INTEGER PRECISION\>
		    \<integer precision value\>
		¦ \<REAL PRECISION\>
		    \<real precision value\>
		¦ \<INDEX PRECISION\>
		    \<index precision value\>
		¦ \<COLOUR PRECISION\>
		    \<colour precision value\>
		¦ \<COLOUR INDEX PRECISION\>
		    \<colour index precision value\>
		{these elements have encoding}
		{dependent parameters }

A.3.3 Picture Descriptor Elements

\<picture descriptor element\>	::=	\<SCALING MODE\>
		    \<scaling spec mode\>
		    \<metric scale factor\>
		¦ \<COLOUR SELECTION MODE\>
		    \<colour select mode\>
		¦ \<LINE WIDTH SPECIFICATION MODE\>
		    \<spec mode\>
		¦ \<MARKER SIZE SPECIFICATION MODE\>
		    \<spec mode\>
		¦ \<EDGE WIDTH SPECIFICATION MODE\>
		    \<spec mode\>

```
                              ¦  <VDC EXTENT>
                                   <point>(2)
                              ¦  <BACKGROUND COLOUR>
                                   <red green blue>
                              ¦  <escape element>
                              ¦  <external element>

<colour select mode>        ::=  <INDEXED>
                              ¦  <DIRECT>

<scaling spec mode>         ::=  <ABSTRACT>
                              ¦  <METRIC>

<metric scale factor>       ::=  <real>

<spec mode>                 ::=  <ABSOLUTE>
                              ¦  <SCALED>

<point>                     ::=  <vdc value>(2)
```

A.3.4 Control Elements

```
<control element>           ::=  <vdc precision>
                              ¦  <AUXILIARY COLOUR>
                                   <colour>
                              ¦  <TRANSPARENCY>
                                   <on-off indicator enumerated>
                              ¦  <CLIP RECTANGLE>
                                   <point>(2)
                              ¦  <CLIP INDICATOR>
                                   <on-off indicator enumerated>

<on-off indicator enumerated> ::=  <ON>
                              ¦  <OFF>

<colour>                    ::=  <colour index>
                              ¦  <red green blue>

<vdc precision>             ::=  <VDC INTEGER PRECISION>
                                   <vdc integer precision value>
                              ¦  <VDC REAL PRECISION>
                                   <vdc real precision value>
                                 {these elements have encoding}
                                 {dependent parameters       }
```

A.3.5 Graphical Elements

```
<graphical element>         ::=  <polypoint element>
                              ¦  <text element>
                              ¦  <cell element>
                              ¦  <gdp element>
                              ¦  <rectangle element>
```

		¦ \<circular element\>
		¦ \<elliptical element\>
\<polypoint element\>	::=	\<POLYLINE\>
		\<point pair\>
		\<point list\>
		¦ \<DISJOINT POLYLINE\>
		\<point pair\>
		\<point pair list\>
		¦ \<POLYMARKER\>
		\<point\>
		\<point list\>
		¦ \<POLYGON\>
		\<point\>(3)
		\<point list\>
		¦ \<POLYGON SET\>
		\<point edge pair\>(3)
		\<point edge pair list\>
\<point list\>	::=	\<point\>*
\<point pair list\>	::=	\<point pair\>*
\<point pair\>	::=	\<point\>(2)
\<point edge pair\>	::=	\<point\>\<edge out flag\>
\<point edge pair list\>	::=	\<point edge pair\>*
\<edge out flag\>	::=	\<INVISIBLE\>
		¦ \<VISIBLE\>
		¦ \<CLOSE_INVISIBLE\>
		¦ \<CLOSE_VISIBLE\>
\<text element\>	::=	\<TEXT\>
		\<point\>
		\<text tail\>
		¦ \<restricted text element\>
\<restricted text element\>	::=	\<RESTRICTED TEXT\>
		\<extent\>
		\<point\>
		\<text tail\>
\<extent\>	::=	\<vdc value\>(2)
\<text tail\>	::=	\<final character list\>
		¦ \<nonfinal character list\>
\<final character list\>	::=	\<FINAL\>
		\<string\>
\<nonfinal character list\>	::=	\<NOT FINAL\>
		\<string\>
		\<char attribute element\>*
		\<spanned text\>

```
<spanned text>              ::= <APPEND TEXT>
                                <text tail>

<cell element>              ::= <CELL ARRAY>
                                <point>(3)
                                <integer>(2)
                                <local colour precision>
                                <colour>(integer1 x integer2)

                                {this element has an encoding}
                                {dependent parameter        }

<local colour precision>    ::= <colour prec value>
                              ¦ <colour index precision value>
                              ¦ <default colour precision indicator>

<gdp element>               ::= <GDP>
                                <gdp identifier>
                                <point list>*
                                <data record>

<gdp identifier>            ::= <integer>

<metafile identifier>       ::= <integer>
<rectangle element>         ::= <RECTANGLE>
                                <point pair>

<circular element>          ::= <CIRCLE>
                                <point>
                                <radius>
                              ¦ <CIRCULAR ARC 3 POINT>
                                <point>(3)
                              ¦ <CIRCULAR ARC 3 POINT CLOSE>
                                <point>(3)
                                <close type>
                              ¦ <CIRCULAR ARC CENTRE>
                                <point>
                                <vdc value>(4)
                                <radius>
                              ¦ <CIRCULAR ARC CENTRE CLOSE>
                                <point>
                                <vdc value>(4)
                                <radius>
                                <close type>

<radius>                    ::= <non-negative vdc value>
<non-negative vdc value>    ::= <vdc value>  {greater or equal to 0}

<close type>                ::= <PIE>
                              ¦ <CHORD>

<elliptical element>        ::= <ELLIPSE>
                                <point>(3)
                              ¦ <ELLIPTICAL ARC>
                                <point>(3)
                                <vdc value>(4)
```

      ¦ <ELLIPTICAL ARC CLOSE>
          <point>(3)
          <vdc value>(4)
          <close type>

A.3.6 Attribute Elements

<primitive attribute element>	::=	<line attribute element>
	¦	<marker attribute element>
	¦	<text attribute element>
	¦	<filled area attribute element>
	¦	<colour table element>
	¦	<aspect source flags>

<line attribute element>	::=	<LINE BUNDLE INDEX>
		<positive index>
	¦	<LINE TYPE>
		<index>
	¦	<LINE WIDTH>
		<size value>
	¦	<LINE COLOUR>
		<colour>

<size value>	::=	<non-negative vdc value>
	¦	<non-negative real>

<non-negative real>	::=	<real> {greater or equal to 0}

<marker attribute element>	::=	<MARKER BUNDLE INDEX>
		<positive index>
	¦	<MARKER TYPE>
		<index>
	¦	<MARKER SIZE>
		<size value>
	¦	<MARKER COLOUR>
		<colour>

<text attribute element>	::=	<char attribute element>
	¦	<string attribute element>

<char attribute element>	::=	<TEXT BUNDLE INDEX>
		<positive index>
	¦	<TEXT FONT INDEX>
		<positive index>
	¦	<CHARACTER EXPANSION FACTOR>
		<real>
	¦	<CHARACTER SPACING>
		<real>
	¦	<TEXT COLOUR>
		<colour>
	¦	<CHARACTER HEIGHT>
		<non-negative vdc value>
	¦	<CHARACTER ORIENTATION>
		<vdc value>(4)

		| <CHARACTER SET INDEX>
		<positive index>
		| <ALTERNATE CHARACTER SET INDEX>
		<positive index>

<string attribute element>   ::= <TEXT PATH>
                                   <path enumerated>
                            | <TEXT PRECISION>
                                   <text precision enumerated>
                            | <TEXT ALIGNMENT>
                                   <horizontal align enumerated>
                                   <vertical align enumerated>
                                   <continuous align value>(2)

<path enumerated>   ::= <RIGHT>
                            | <LEFT>
                            | <UP>
                            | <DOWN>

<text precision enumerated> ::= <STRING>
                            | <CHARACTER>
                            | <STROKE>

<horizontal align enumerated> ::= <NORMAL HORIZONTAL>
                            | <LEFT>
                            | <CENTRE>
                            | <RIGHT>
                            | <CONTINUOUS HORIZONTAL>

<vertical align enumerated> ::= <NORMAL VERTICAL>
                            | <TOP>
                            | <CAP>
                            | <HALF>
                            | <BASE>
                            | <BOTTOM>
                            | <CONTINUOUS VERTICAL>

<continuous align value>   ::= <real>

<filled area attribute element> ::= <FILL BUNDLE INDEX>
                                   <positive index>
                            | <INTERIOR STYLE>
                                   <interior enumerated>
                            | <FILL COLOUR>
                                   <colour>
                            | <HATCH INDEX>
                                   <index>
                            | <PATTERN INDEX>
                                   <positive index>
                            | <EDGE BUNDLE INDEX>
                                   <positive index>
                            | <EDGE TYPE>
                                   <index>
                            | <EDGE WIDTH>
                                   <size value>
                            | <EDGE COLOUR>

```
                              <colour>
                          ¦ <EDGE VISIBILITY>
                              <on-off indicator enumerated>
                          ¦ <FILL REFERENCE POINT>
                              <point>
                          ¦ <PATTERN TABLE>
                              <positive index>
                              <integer>(2)
                              <local colour precision>
                              <colour>(integer1 x integer2)
                              {this element has an encoding}
                              {dependent parameter        }
                          ¦ <PATTERN SIZE>
                              <vdc value>(4)
```

`<interior enumerated>`	`::= <HOLLOW>`
	`¦ <SOLID>`
	`¦ <PATTERN>`
	`¦ <HATCH>`
	`¦ <EMPTY>`

`<colour table element>`	`::= <COLOUR TABLE>`
	`    <starting index>`
	`    <red green blue>+`

`<starting index>`	`::= <colour index>`
`<aspect source flags>`	`::= <ASPECT SOURCE FLAGS>` `    <asf pair>+`
`<asf pair>`	`::= <asf type>` `    <asf>`

`<asf type>`	`::= <LINE TYPE ASF>`
	`¦ <LINE WIDTH ASF>`
	`¦ <LINE COLOUR ASF>`
	`¦ <MARKER TYPE ASF>`
	`¦ <MARKER SIZE ASF>`
	`¦ <MARKER COLOUR ASF>`
	`¦ <TEXT FONT ASF>`
	`¦ <TEXT PRECISION ASF>`
	`¦ <CHARACTER EXPANSION FACTOR ASF>`
	`¦ <CHARACTER SPACING ASF>`
	`¦ <TEXT COLOUR ASF>`
	`¦ <INTERIOR STYLE ASF>`
	`¦ <FILL COLOUR ASF>`
	`¦ <HATCH INDEX ASF>`
	`¦ <PATTERN INDEX ASF>`
	`¦ <EDGE TYPE ASF>`
	`¦ <EDGE WIDTH ASF>`
	`¦ <EDGE COLOUR ASF>`

`<asf>`	`::= <INDIVIDUAL>` `¦ <BUNDLED>`

A.3.7 Escape Elements

\<escape element\>	::=	\<ESCAPE\>
		\<identifier\>
		\<data record\>
\<identifier\>	::=	\<integer\>

A.3.8 External Elements

\<external element\>	::=	\<MESSAGE\>
		\<action flag\>
		\<string\>
	|	\<APPLICATION DATA\>
		\<identifier\>
		\<data record\>
\<action flag\>	::=	\<YES\>
	|	\<NO\>

A.4 Terminal Symbols

The following are the terminals in this grammar. Their representation is dependent on the encoding scheme used. In annex A of the subsequent parts of this Standard, these encoding-dependent symbols are further described.

    \<element name\>
    \<integer\>
    \<real\>
    \<vdc value\>
    \<string\>
    \<colour index\>
    \<red green blue\>
    \<integer precision value\>
    \<real precision value\>
    \<index precision value\>
    \<colour precision value\>
    \<colour index precision value\>
    \<default colour precision indicator\>
    \<vdc integer precision value\>
    \<vdc real precision value\>
    \<colour list\>
    \<data record\>

The CGM opcodes are encoding dependent. A complete list of them can be found in the productions for \<element name enumerated\> below.

The enumerated types:

    \<INTEGER\>
    \<REAL\>

<ON>
<OFF>
<INDEXED>
<DIRECT>
<ABSTRACT>
<METRIC>
<ABSOLUTE>
<SCALED>
<94 CHAR>
<96 CHAR>
<MULTI-BYTE 94 CHAR>
<MULTI-BYTE 96 CHAR>
<COMPLETE CODE>
<BASIC 7-BIT>
<BASIC 8-BIT>
<EXTENDED 7-BIT>
<EXTENDED 8-BIT>
<INVISIBLE>
<VISIBLE>
<CLOSE_INVISIBLE>
<CLOSE_VISIBLE>
<PIE>
<CHORD>
<FINAL>
<NOT FINAL>
<INDIVIDUAL>
<BUNDLED>
<HOLLOW>
<SOLID>
<PATTERN>
<HATCH>
<EMPTY>
<STRING>
<CHARACTER>
<STROKE>
<RIGHT>
<LEFT>
<UP>
<DOWN>
<NORMAL HORIZONTAL>
<CENTRE>
<CONTINUOUS HORIZONTAL>
<NORMAL VERTICAL>
<TOP>
<CAP>
<HALF>
<BASE>
<BOTTOM>
<CONTINUOUS VERTICAL>
<YES>
<NO>
<LINE TYPE ASF>
<LINE WIDTH ASF>
<LINE COLOUR ASF>
<MARKER TYPE ASF>
<MARKER SIZE ASF>

<MARKER COLOUR ASF>
<TEXT FONT ASF>
<TEXT PRECISION ASF>
<CHARACTER EXPANSION FACTOR ASF>
<CHARACTER SPACING ASF>
<TEXT COLOUR ASF>
<INTERIOR STYLE ASF>
<HATCH INDEX ASF>
<PATTERN INDEX ASF>
<FILL COLOUR ASF>
<EDGE TYPE ASF>
<EDGE WIDTH ASF>
<EDGE COLOUR ASF>

<element name enumerated>   ::= <BEGIN METAFILE>
                             | <END METAFILE>
                             | <BEGIN PICTURE>
                             | <BEGIN PICTURE BODY>
                             | <END PICTURE>
                             | <METAFILE VERSION>
                             | <METAFILE DESCRIPTION>
                             | <VDC TYPE>
                             | <INTEGER PRECISION>
                             | <REAL PRECISION>
                             | <INDEX PRECISION>
                             | <COLOUR PRECISION>
                             | <COLOUR INDEX PRECISION>
                             | <MAXIMUM COLOUR INDEX>
                             | <COLOUR VALUE EXTENT>
                             | <METAFILE ELEMENT LIST>
                             | <METAFILE DEFAULTS REPLACEMENT>
                             | <FONT LIST>
                             | <CHARACTER SET LIST>
                             | <CHARACTER CODING ANNOUNCER>
                             | <SCALING MODE>
                             | <COLOUR SELECTION MODE>
                             | <LINE WIDTH SPECIFICATION MODE>
                             | <MARKER SIZE SPECIFICATION MODE>
                             | <EDGE WIDTH SPECIFICATION MODE>
                             | <VDC EXTENT>
                             | <BACKGROUND COLOUR>
                             | <VDC INTEGER PRECISION>
                             | <VDC REAL PRECISION>
                             | <AUXILIARY COLOUR>
                             | <TRANSPARENCY>
                             | <CLIP RECTANGLE>
                             | <CLIP INDICATOR>
                             | <POLYLINE>
                             | <DISJOINT POLYLINE>
                             | <POLYMARKER>
                             | <TEXT>
                             | <RESTRICTED TEXT>
                             | <APPEND TEXT>
                             | <POLYGON>
                             | <POLYGON SET>
                             | <CELL ARRAY>

| <GDP>
| <RECTANGLE>
| <CIRCLE>
| <CIRCULAR ARC 3 POINT>
| <CIRCULAR ARC 3 POINT CLOSE>
| <CIRCULAR ARC CENTRE>
| <CIRCULAR ARC CENTRE CLOSE>
| <ELLIPSE>
| <ELLIPTICAL ARC>
| <ELLIPTICAL ARC CLOSE>
| <LINE BUNDLE INDEX>
| <LINE TYPE>
| <LINE WIDTH>
| <LINE COLOUR>
| <MARKER BUNDLE INDEX>
| <MARKER TYPE>
| <MARKER SIZE>
| <MARKER COLOUR>
| <TEXT BUNDLE INDEX>
| <TEXT FONT INDEX>
| <TEXT PRECISION>
| <CHARACTER EXPANSION FACTOR>
| <CHARACTER SPACING>
| <TEXT COLOUR>
| <CHARACTER HEIGHT>
| <CHARACTER ORIENTATION>
| <TEXT PATH>
| <TEXT ALIGNMENT>
| <CHARACTER SET INDEX>
| <ALTERNATE CHARACTER SET INDEX>
| <FILL BUNDLE INDEX>
| <INTERIOR STYLE>
| <FILL COLOUR>
| <HATCH INDEX>
| <PATTERN INDEX>
| <EDGE BUNDLE INDEX>
| <EDGE TYPE>
| <EDGE WIDTH>
| <EDGE COLOUR>
| <EDGE VISIBILITY>
| <FILL REFERENCE POINT>
| <PATTERN TABLE>
| <PATTERN SIZE>
| <COLOUR TABLE>
| <ASPECT SOURCE FLAGS>
| <ESCAPE>
| <MESSAGE>
| <APPLICATION DATA>
| <DRAWING SET>
| <DRAWING PLUS CONTROL SET>

B Guidelines for Private Encodings

NOTE - This annex is not part of the Standard; it is included for information purposes only.

A functionally conforming metafile (see 7.2) may use a non-standardized (private) encoding. For such a encoding to be a candidate for standardization, it should conform to the following conditions.

a) All CGM elements shall have a specified encoding, with the exception of the precision commands, which may not be applicable to a particular encoding. An element that sets an interpretation mode for other elements may be implicit in the commands that it affects, as opposed to being coded as a separate element. (For example, a procedural encoding might include separate calls for LINE_COLOUR_DIRECT(R,G,B) and LINE_COLOUR_INDEX(I) and omit COLOUR SELECTION MODE.)

b) All CGM functionality shall be realizable (for example, both integer and real coordinates), except where noted above under (a).

c) The encoding shall utilize sufficient precision to accomodate the Minimum Suggested Capability List (see D.5). For example, the representation of bundle indices shall be able to represent the range 1..5, inclusive.

Furthermore, in keeping with the design guidelines used for developing the standardized encodings, it is suggested that designers of private encodings ensure:

— the ability to translate a metafile encoded in one of the standardized encodings into the private encoding;

— the corresponding ability to translate a privately encoded metafile to one of the standardized bindings.

These requirements should be considered as recommendations for those designing non-standardized encodings. In addition, it is strongly recommended that encodings support the range of coordinate data precisions standardized in part 3, Binary Encoding.

C Reference Models

NOTE - This annex is not part of the Standard; it is included for information purposes
only.

The CGM is designed to be usable by and useful to a wide range of applications, graphics systems, and
devices or workstations. The figures in this annex are context diagrams — they illustrate the relationship of the CGM to the other components of a graphics system. They are not meant to be detailed
specifications of interfaces and procedures, but rather conceptual illustrations of relationships.

Another annex of this Standard, annex E, presents a detailed model of the relationship of the CGM to
the GKS standard (ISO 7942).

CGM is graphics-system independent. Figure 16 shows the CGM in the context of a generic device-
independent graphics system. The metafile generator exists below and is invoked by the application-
callable layer, at the level of the device or workstation driver. The metafile generator records device
independent picture descriptions, conceptually in parallel with the presentation of images on actual devices.

By design, the CGM may be interpreted either by a special process that does not involve the high-level
facilities of a general purpose graphics system (figure 17), or may be interpreted using the application-
callable services of a device independent graphics package (figure 18). Figure 17 might illustrate, for
example, a scenario where a CGM has been customized to a particular application and device environment. It might also illustrate the case of a target graphics device which has the CGM element set in
hardware or firmware. Figure 18 illustrates an easy and convenient way to render the pictorial content
of a metafile in a more device-independent and application-independent manner.

A GKS implementation could be the device-independent graphics system in these context diagrams.
The GKS standard, however, specifies the relationship of GKS to metafiles with somewhat more detail
than these diagrams contain. See annex E for further discussion of the CGM/GKS relationship.

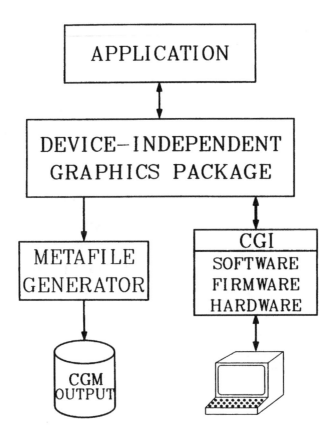

Figure 16. Relationship of CGM to traditional graphics package.

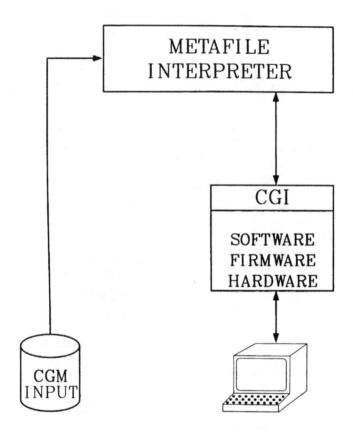

Figure 17. Metafile interpretation with no graphics package.

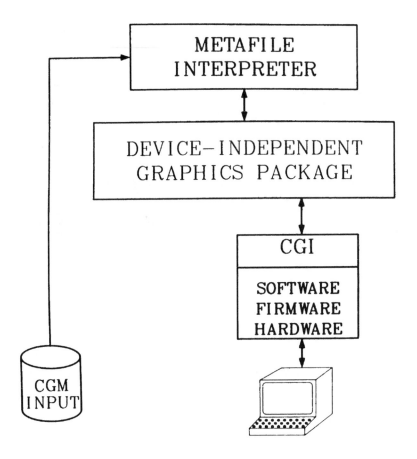

Figure 18. Metafile interpretation using traditional graphics package.

D Guidelines for Metafile Generators and Interpreters

NOTE - This annex is not part of the Standard; it is included for information purposes only.

D.1 Introduction

The CGM standardizes the contents, syntax, and semantics of a set of CGM elements. It does not standardize the metafile interpreter. In some situations, it may not be possible for a metafile interpreter to accurately render the contents of a metafile. These situations include cases where the interpreter lacks functionality to render correctly specified elements in the metafile, as well as cases where the contents of the metafile are improperly specified (errors and degeneracies in the metafile data).

In some closed environments, for example generation and interpretation of metafiles by a GKS system, the environment itself may dictate what is to be done in some or all of these situations. In other environments there may not be any rules or guidelines for implementors to follow. For predictability and uniformity of results, it is thus useful to suggest a common approach to situations in which a metafile interpreter cannot accurately render the contents of a metafile, and in which other criteria for deciding a response are lacking.

$D.2$ deals with errors and degeneracies at a general level. $D.3$ and $D.4$ contain recommended approximations for the interpretation of CGM elements where no one-to-one mapping exists between a CGM element and display device capability. These sub-clauses also deal with some particular mathematical degeneracies and ambiguities. $D.5$ lists minimum suggested capabilities for CGM interpreters.

D.2 Errors and Degeneracies

As detailed in 5.1, three categories of degeneracies are identified in the specification of metafile elements:

a) syntactic errors;

b) geometrically degenerate primitives;

c) mathematical singularities and ambiguities.

Subsequent sub-clauses of this annex contain some recommendations for implementors of interpreters for dealing with such exception conditions in metafile contents. Regardless of the strategy chosen by an implementor, it is recommended that it be documented for users of metafile interpreters.

D.2.1 Syntax Errors

The general recommendation to interpreters regarding syntax errors is to recover as much information as possible, and if there is a reasonable interpretation of the element (particularly in the case of primitives) to generate some visible output — "do the best you can".

Some syntax errors may make it impossible to meaningfully parse the remainder of a metafile, e.g., a data count does not correspond to the amount of data actually encoded in the element. This class of syntax error is particularly sensitive to the specific encoding.

Other types of syntax error are less severe, in that an interpreter is able to continue parsing the metafile. In some cases assumptions can be made and some meaningful output usually generated. In other cases, e.g., invalid values in attribute setting elements, there may be an infinite number of equally likely possibilities with nothing to recommend between them, and it is suggested that the element be ignored.

Examples:

DISJOINT POLYLINE

If the number of points is odd and greater than 1, it is suggested that the interpreter assume that the last point is the odd one, and display the primitive derived by discarding the last point.

CELL ARRAY

If nx or ny is zero in the cell array specification, it is suggested that no output be generated by the metafile interpreter.

D.2.2 Geometrically Degenerate Primitives

Zero-length and zero-area specifications comprise the majority of such degeneracies. Typical examples of such specifications are a POLYLINE element in which all data points are coincident (zero length), a POLYGON element where all of the data points alternate between two distinct points (zero area), or a CIRCLE element with zero radius (zero area).

It is recommended in most cases that some visible output be generated for geometrically degenerate primitives. In some cases the application may have reason to render the degeneracy in a particular way. For example, there may be a style that is a natural or "continuous" limiting appearance of the primitive as it approaches degeneracy. In cases in which the application has no strong reasons to pick a particular rendering, it is suggested that the recommendations below be followed.

D.2.2.1 Zero Length. Zero-length degeneracies apply to the line elements (see 4.6 for a list of line elements). When the specification of a line element degenerates to zero length, it is recommended that the interpreter display a dot in the current line colour and at the size of the current line width. See *D.*3 and *D.*4 for interpreter fallback in the case that these attributes cannot be exactly honoured.

D.2.2.2 Zero Area. Zero-area degeneracies apply to the filled-area elements (see 4.6 for a list of filled-area elements). Two subcategories are recognized — the primitive degenerates to a dot or the primitive degenerates to a line. It is recommended that the dot or line be displayed with the FILL COLOUR if EDGE VISIBILITY is 'off', unless INTERIOR STYLE is 'empty'. If EDGE VISIBILITY is 'on' the dot or line is displayed with the edge attributes.

A CELL ARRAY element whose 3 points define a zero-area parallelogram falls in this category as well, except that it is always treated as if EDGE VISIBILITY were 'on'.

D.3 General Guidelines

D.3.1 Indexes

With the exception of colour, an out-of-range index in an indexed selection element causes selection of the default index value. Out-of-range colour selection indexes are mapped to supported colour indexes in a deterministic fashion; implementors of interpreters should document the mapping. An out-of-range index on an index definition element (for example, COLOUR TABLE) is ignored.

D.3.2 Colour Model

There are two key assumptions to the formation of direct and indexed colour specifications:

a) Many computer graphics devices are inherently indexed in regard to colour attribute selection. These devices include the following:

    1) frame buffer video bit maps (index is pixel value)

    2) pen plotters (index is pen identifier)

    3) stroke devices (index is analog of voltage for refresh intensity or beam penetration).

b) All of these devices map these indexes to visible attributes through a map. These maps are either loadable under program control (type 1 above), operator control (2), or are fixed in hardware (3) in a way encodable in software at some level.

The most direct hardware control over these devices is, thus, by colour selection by index. Direct hardware control (by index control) presumes that software need have enough knowledge of the accessible hardware range to be able to control the device usefully.

To utilize direct colour specification (if, by assumption a, devices implement indexed specifiers), we need to reverse-map the map noted in assumption b.

A "closest match" element between the directly specified colour and the entries of the table could be based on several algorithms. For example:

1) Minimum spatial distance, as computed in the RGB colour cube.

2) Component-by-component comparison by minimum weight XORing. In fixed tables, colours tend to be evenly distributed throughout the colour space, or at least across a plane through the space. This aids the closest match problem, because there is likely to be a colour somewhat near the target.

In loadable tables of indeterminate length, it may be advisable to load colours from disparate points in the colour space early in the table before the colour selection table is exhausted.

Only the COLOUR TABLE element and the BACKGROUND COLOUR element can change the colour map (if it exists). Direct specification requires that closest match be used to protect static attributes in a display, which, for portability, are the default.

D.3.2.1 Mapping to Monochrome. In American colour television systems (NTSC encoding), colour signals are translated to grey scale for monochrome reception by the following equation:

$$Y = 0.30R + 0.59G + 0.11B$$

where R, G, and B are the intensity values of the red, green, and blue components, and Y is the resulting luminance value. This mapping is suggested for metafile interpreters.

The integer mapping of this equation would be, approximately

$$Y' = 3R + 6G + 1B.$$

D.3.2.2 Specified Versus Realized Colour Precision. It is likely that the numerical ranges of the RGB colour components in a metafile (as specified by the COLOUR VALUE EXTENT element) will differ from the component ranges available on a device. If the metafile was generated with different component ranges than available on the device, the metafile components are mapped into the device components by linearly mapping the metafile component ranges onto the device component ranges.

Figure 21 gives an example of such a mapping. The metafile generator declares a COLOUR PRECISION of 3 bits and a COLOUR VALUE EXTENT of (0,0,0), (7,7,7). The devices has 4 bits of precision in each colour component, and a component range of 0-15. The example shows only the mapping for the red component of colour — green and blue map exactly the same way. The metafile COLOUR VALUE EXTENT effectively states that there are eight levels of red expressable in the metafile — evenly spaced from no red component to full red. The device precision of four bits says there are 16 levels provided by the device. The mapping preserves full intensity and zero intensity of the components. Note that this discussion of direct colour specification also applies to the direct colour values used to set colour table entries.

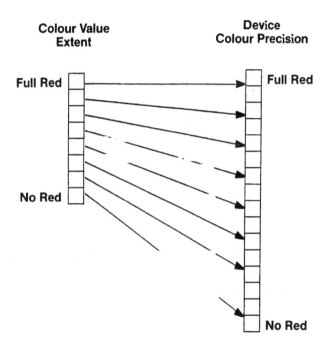

Figure 19. Mapping of direct colour specification from metafile to device.

D.4 Guidelines Specific to Element Classes

D.4.1 Delimiter Elements

It is intended that the BEGIN PICTURE BODY element clear the view surface. This is necessary if pictures are randomly accessed. If the metafile interpreter is composing an image from multiple CGM pictures, the interpreter will clear the view surface only before the first picture in the image.

BEGIN PICTURE BODY
> The element typically causes the view surface to be cleared to the background colour, as specified by the BACKGROUND COLOUR element if present in the Picture Descriptor or default background colour if not.

END PICTURE
> It is suggested that interpretation of the END PICTURE element guarantee that the picture reflect the display of all the elements in the picture body.

D.4.2 Metafile Descriptor Elements

There are no interpreter guidelines for any of the Metafile Descriptor Elements.

D.4.3 Picture Descriptor Elements

There are no interpreter guidelines for any of the Picture Descriptor Elements.

D.4.4 Control Elements

AUXILIARY COLOUR
> This element is intended to address hardware features commonly available in raster display devices. Some devices may have no such capabilities, or may have a subset of the capabilities to which this element pertains. Simulation of such a feature may be very complex and expensive. It is not intended that an interpreter need simulate the feature when it is not available in the hardware or firmware.

CLIP RECTANGLE
> If any part of the CLIP RECTANGLE is outside VDC EXTENT, clipping should occur at the intersection of CLIP RECTANGLE and VDC EXTENT.

D.4.5 Graphical Primitive Elements

APPEND TEXT
> It is may be difficult for an implementation to buffer text attribute changes which are permitted between partial text elements. The suggested fallback behaviour for interpreters is to accept the attribute change, accumulate and concatenate the pieces of the string, and align and display the completed text string with the set of attributes in effect at the time the string is completed (an APPEND TEXT with a 'final' flag).

RESTRICTED TEXT
> Some fonts may have kerns extending outside of the character body, either horizontally or vertically. For the RESTRICTED TEXT element, which requires all visible portions of a character symbol to be confined to a parallelogram derived from the element's extent parameters, such kerns are treated in an implementation-dependent way. One way is to treat the extent parallelogram as a clipping boundary as far as kerns are concerned. Another way is for the interpreter to make "worst case" assumptions about the font, and process the text string as if there were an extra half-space or space on either end to allow for kerns — this way insures that all portions of the text are drawn and within the parallelogram.

CELL ARRAY
It is suggested that a device that cannot display the CELL ARRAY element draw a parallelogram corresponding to the specified area. The parallelogram is drawn according to the filled-area edge attributes, with the exception that EDGE VISIBILITY is ignored.

If the three points defining the CELL ARRAY form a parallelogram and the CELL ARRAY cannot be displayed as a parallelogram, then it is implementation dependent whether the specified parallelogram is drawn or the CELL ARRAY is displayed without skewing.

CIRCULAR ARC 3 POINT
A dot is drawn for an ARC with only one distinct point.

If the element has only two distinct points:

— If the intermediate point coincides with the starting point, a line is drawn between the starting point and the ending point.

— If the intermediate point coincides with the ending point, a line is drawn between the starting point and the ending point.

— If the starting point coincides with the ending point, a circle is drawn such that the line from the starting point to the intermediate point is a diameter of the circle.

If the element has three collinear coordinates:

— If the intermediate point lies between the starting point and ending point, on the line through those two points, then a line segment is drawn from the starting point to the ending point.

— If the intermediate point does not lie between the starting point and the ending point, but is on the line through those two points, then two different semi-infinite lines are drawn: one begins at the starting point and extends in the direction of the vector from the ending point to the starting point; the other begins at the ending point and extends in the direction of the vector from the starting point to the ending point.

CIRCULAR ARC 3 POINT CLOSE
In the case that the starting and ending points coincide and the intermediate point is distinct, the interior of the defined circle (see CIRCULAR ARC 3 POINT) is rendered according to the filled-area interior attributes, and the edge is rendered according to the edge attributes. When in addition the 'close type' is 'pie', as well as filling the circle a radius should also be drawn with the current edge attributes.

In all the other cases (see CIRCULAR ARC 3 POINT) the defined arc does not have a unique or finite centre, and an area (e.g., half-plane) cannot be uniquely determined. Therefore in these cases the boundary/edge is generated as specified for each of the singular conditions described above, under the guidelines for CIRCULAR ARC 3 POINT. The suggestions for rendering of the boundary and/or edge of zero-area primitives (see D.2) should be applied to the drawing of the specified lines in these cases.

CIRCULAR ARC CENTRE
If the start ray and the end ray coincide, it is recommended that the interpreter draw the full circle.

CIRCULAR ARC CENTRE CLOSE
If the start ray and the end ray coincide, it is recommended that the interpreter draw and fill the full circle. When in addition the 'close type' is 'pie', as well as filling the circle a radius should also be drawn with the current edge attributes.

Elliptical Elements (rendering techniques)
There are a number of techniques for rendering an ellipse specified by the parameterization of this Standard. One such technique involves the use of a pair of equations of a single parameter (t) whose coefficients may derived from the conjugate diameters. One equation

generates x-coordinates and the other generates y-coordinates as t varies. This technique and its use are included in claims covered by patents which are held by Conographic Corporation and which are applicable in many ISO member nations; by publication of this Standard, no position is taken with respect to the validity of this claim or of any patent rights in connection therewith.

Elliptical Elements (supplementary information)

To supplement the syntactic definition of elliptical elements in clauses 4 and 5 of this part of the Standard, a mathematical representation of ellipses is given here.

An ellipse is parameterized by its centrepoint and an endpoint from each of the diameters of any CDP. The following equations define the ellipse in terms of these data. For simplicity, the equations are presented for an ellipse centred at the origin. Let $P1=(x_1,y_1)$ and $P2=(x_2,y_2)$ be the two CDP endpoints. The equation for the ellipse is:

$$\frac{(x\cos(\alpha) + y\sin(\alpha))^2}{a^2} + \frac{(x\sin(\alpha) - y\cos(\alpha))^2}{b^2} = 1$$

where

$$a = \frac{y_u - y_2}{\sin(\gamma)}$$

$$b = \frac{y_2 - y_v}{\sin(\gamma)}$$

$$\gamma = arctan\left(\frac{y_c - y_2}{x_2 - x_c}\right)$$

$$\alpha = arctan\left(-\frac{y_v}{x_v}\right)$$

$$x_c = \frac{x_2 + y_1}{2}$$

$$y_c = \frac{y_2 - x_1}{2}$$

$$y_u = y_c + d\sin(\gamma)$$

$$x_v = x_c + d\cos(\gamma)$$

$$y_v = y_c - d\sin(\gamma)$$

$$d = \frac{x_c}{\cos(\phi)}$$

$$\phi = arctan\left(\frac{y_c}{x_c}\right)$$

The generalization for an ellipse with an arbitrary centrepoint $M=(x_m,y_m)$ is straightforward.

ELLIPTICAL ARC

If the start ray and end ray are coincident, it is recommended that the interpreter draw the full ellipse.

ELLIPTICAL ARC CLOSE
>   If the start ray and end ray are coincident, it is recommended that the interpreter draw and
>   fill the full ellipse. When in addition the 'close type' is 'pie', as well as filling the ellipse a
>   radius should also be drawn with the current edge attributes.

D.4.6 Attribute Elements

BUNDLES
>   Clause 4 describes the component attributes comprising the various attribute bundles. It is
>   anticipated that future revisions of this Standard will include settable bundles. To avoid
>   possible conflict with future revisions of this Standard, it is strongly recommended that inter-
>   preters use only these components to achieve distinguishability and not use other, currently
>   nonstandardized attributes. If, however, it is not feasible for the interpreter to adhere to
>   this guideline, then the use of device-dependent, nonstandardized attributes (for example,
>   blink or highlight) is a reasonable alternative.

LINE TYPE
>   If the specified implementation-dependent line type is not available, 'solid' is used.
>
>   If line type cannot be maintained continuously across interior vertices of a single polyline ele-
>   ment, then restarting the line pattern at each interior vertex is the recommended action.

LINE WIDTH
>   If a device cannot produce a line of the exact specified width, the closest implemented width
>   is chosen.

MARKER TYPE
>   If the specified implementation-dependent marker type is not available, the marker type
>   'asterisk' is used.

MARKER SIZE
>   The marker size is mapped to the nearest available marker size on the device. The effect of
>   MARKER SIZE on implementation-dependent markers is implementation dependent.

TEXT PRECISION
>   If a specified text precision is not available, the next more precise implemented text precision
>   is chosen. If no such (more precise) value is available, it is interpreter dependent whether:
>   (1) the next lower value is used, or (2) whether a substitute font will be invoked temporarily
>   to provide the requested text precision. Note that the font may already be a temporary
>   invocation to provide a character set not provided by the current TEXT FONT INDEX (see
>   CHARACTER SET INDEX in this annex).

CHARACTER EXPANSION FACTOR
>   The next available value smaller than or equal to the specified value is selected. If no such
>   value is available, the next larger value is selected. The effective character height and char-
>   acter width are set to values that allow characters to completely fit into an enclosing rectan-
>   gle determined by the desired character height and width. If this is not possible, the smallest
>   available character size is used.

CHARACTER SPACING
>   The next available value smaller than or equal to the specified value is selected. If no such
>   value is available, the next larger value is selected.

CHARACTER HEIGHT
>   The next available value smaller than or equal to the specified value is selected. If no such
>   value is available, the next larger value is selected.

CHARACTER ORIENTATION
>   If the specified character up vector is not available, the nearest available vector is chosen. If
>   two are equally near, the one in a positive angular direction is chosen. When the character
>   up vector and the character base vector are not at right angles, and hardware text is being

used which cannot be skewed, the base vector is used to determine character orientation. If the character path is 'left' or 'right', the character base vector determines placement of character origins; if the path is 'up' or 'down', the character up vector determines placement of character origin.

TEXT PATH

The fallback value for 'left' is 'right', the fallback value for 'up' is 'down', and vice versa. If the fallback value recommended above is not available, 'right' is chosen.

TEXT ALIGNMENT

For the TEXT element, if an alignment value is not available, the closest available value is used. For RESTRICTED TEXT, the interpreter should honour alignment exactly for the positioning of the extent parallelogram. In cases where multiple precisions are specified within a complete string, the interpreter may, if necessary, use a precision lower than the highest overall precision for alignment purposes.

CHARACTER SET INDEX

If the selected character set is not available in the selected font at the time of elaboration of TEXT, APPEND TEXT, or RESTRICTED TEXT elements, the font will be temporarily changed to one where the selected character set can be represented.

FONT DESIGN AND THE FONT COORDINATE SYSTEM

Sufficient white space should be allowed when setting the limits of the character body relative to the characters in the font to permit characters to be displayed with their bodies flush without producing conflicts or overlaps between ascenders and descenders, and with normal spacing between characters. This permits standardized use of the continuous alignment parameter in TEXT alignment, and supports the intent of spacing with CHARACTER SPACING = 0.0.

INTERIOR STYLE

If the requested interior style is not available, 'hollow' is used.

HATCH INDEX

The ideal value of 'positive slope' is an angle of positive 45 degrees, and of 'negative slope' is an angle of positive 135 degrees (negative 45 degrees). If the interperter cannot render these lines with these slopes, then lines with slopes that are recognizably similar may be used. Angles of 30-60 degrees and 120-150 degrees are reasonable approximations respectively for 'positive slope' and 'negative slope'.

EDGE TYPE

If the specified implementation-dependent edge type is not available, 'solid' is used.

If edge type cannot be maintained continuously across interior vertices of a single POLYGON or POLYGON SET element, then restarting the line pattern at each interior vertex is the recommended action.

EDGE WIDTH

If a device cannot produce a edge of the exact specified width, the closest implemented width is chosen.

This Standard does not mandate the alignment (i.e., centring) of the finite-width realized edge and the zero-width ideal edge of a filled-area element. It is suggested that applications use one of the following two alignments, according to the needs of the application:

1)   The realized edge is centred on the ideal edge. This alignment is most straightforward to implement, but has the disadvantage that the complete displayed filled-area element extends outside of the ideal edge.

2)   The realized edge is everywhere inside and everywhere contiguous to the ideal edge. This alignment has the advantage that the complete displayed filled-area occupies exactly the area defined by the ideal edge. It is more difficult to implement than the first alternative.

PATTERN SIZE

    If a device cannot produce skewed and/or rotated patterns, then the pattern size is interpreted as if the pattern height vector were vertical and the pattern width vector were horizontal. If a device cannot produce a pattern of the exact specified size, the closest implemented size is chosen.

ASPECT SOURCE FLAGS

    If the initial ASPs are not altered, the expected behaviour of the interpreter is

    a)   as if individual specification of bundled aspects were not a system feature, if the initial values of all the ASFs are 'bundled'; or

    b)   as if specification of bundled aspects via a bundle were not a system feature, if the initial values of all the ASFs are 'individual'.

D.4.7 Escape Elements

There are no interpreter guidelines for Escape Elements.

D.4.8 External Elements

There are no interpreter guidelines for External Elements.

D.5 Minimum Suggested Capability List

For uniformity of results when interpreting a metafile, it is suggested that CGM interpreters have at least the following capabilities.

Capability	Minimum Suggested Interpreter Support
CHARACTER CODING ANNOUNCER	Basic 7-bit
FONT LIST	at least one font capable of displaying the character set described below, see CHARACTER SET LIST
CHARACTER SET LIST	at least one character set which includes the nationality-independent subset of ISO 646 in the positions specified in ISO 646
BACKGROUND COLOUR	1, interpreter dependent
AUXILIARY COLOUR	transparent
TRANSPARENCY	on
TEXT string size for alignment	80 characters
Vertices for POLYGON and POLYGON SET	128
LINE BUNDLE INDEX	5
LINE TYPE	solid,dash,dot,dash-dot,dash-dot-dot
LINE WIDTH	1, interpreter dependent
LINE COLOUR	1, interpreter dependent
MARKER BUNDLE INDEX	5
MARKER TYPE	dot,plus,asterisk,circle,cross
MARKER SIZE	1, interpreter dependent
MARKER COLOUR	1, interpreter dependent
TEXT BUNDLE INDEX	2
TEXT FONT INDEX	1
TEXT PRECISION	string,character
CHARACTER EXPANSION FACTOR	1, interpreter dependent
CHARACTER SPACING	1, interpreter dependent
TEXT COLOUR	1, interpreter dependent
CHARACTER HEIGHT	1, interpreter dependent
CHARACTER ORIENTATION	along the axes of VDC space
TEXT PATH	right,left,up,down
TEXT ALIGNMENT	normal vertical, top, bottom, baseline, normal horizontal, left, centre, right
CHARACTER SET INDEX	1
ALTERNATE CHARACTER SET INDEX	1
FILL BUNDLE INDEX	5
INTERIOR STYLE	hollow,solid,pattern,hatch,empty
FILL COLOUR	1, interpreter dependent
HATCH INDEX	1, interpreter dependent
PATTERN INDEX	1, interpreter dependent
EDGE BUNDLE INDEX	5
EDGE COLOUR	Same as LINE COLOUR
EDGE TYPE	Same as LINE TYPE
EDGE WIDTH	Same as LINE WIDTH
PATTERN SIZE	1, interpreter dependent

E Relationship of CGM and GKS

NOTE - This annex is not part of the Standard; it is included for information purposes only.

E.1 Introduction

The GKS standard (ISO 7942) includes the concepts of metafile input and output workstations, as well as functions providing access to and interpretation of metafiles. It does not, however, contain a metafile definition as part of the standard. Annex E of the standard contains the definition of a "GKS Metafile" (GKSM). The suggested GKSM is basically an audit trail at the workstation interface of the GKS functions, containing all of the dynamic information originally available in the output stream of an entire interactive graphics session.

This CGM Standard defines a metafile for the capture of static picture definitions. Specifically, it contains adequate functionality to serve as a picture capture mechanism in the GKS environment. Since use of the CGM was not intended to be restricted to GKS environments, there is not a one-to-one mapping between the functions of the two standards — CGM lacks some GKS facilities while offering others not available in GKS. This raises some questions about generation and interpretation of CGM in a GKS environment.

Because neither the CGM nor the GKS standardizes the relationship of graphics system to metafile, there is no unique useful relationship between the two. This annex presents one relationship that is suitable for defining the use of CGM in GKS environments.

E.2 Scope

While the CGM can capture static picture definitions from any level of GKS, the relationship of CGM elements to GKS functions is most straightforward at level 0 of GKS. When the dynamic functions from levels 1 and 2 of GKS are used by applications, the strategies for generating proper picture definitions are both more numerous and complex. The best strategy to use in given circumstances is dictated by implementation and application requirements. Hence, this annex presents a detailed mapping between GKS functions and CGM elements only for functions at level 0 of GKS.

The scope of this annex is further limited to generation of metafiles by GKS, and interpretation of GKS-generated metafiles by GKS facilities. There are many other scenarios for generation and interpretation of the CGM, such as interpretation by GKS of metafiles not generated by GKS and interpretation by non-GKS processes of GKS-generated metafiles. These scenarios are not dealt with in this annex. Annex C presents context models dealing with such cases.

E.3 Overview of the Differences Between GKS and CGM

While CGM supports all of the basic output functionality of GKS, a one-to-one mapping between GKS and CGM is not possible in all cases. Specifically:

1) Some GKS functions have no counterpart as CGM elements:

   a) Elements for setting or changing the workstation transformation.

   b) The 'SET xxx REPRESENTATION' functions (xxx = POLYLINE, POLYMARKER, TEXT, or FILL AREA) for setting workstation attributes in the corresponding bundle tables.

   c) All functions referring to segments.

2) Some CGM elements have no counterpart as GKS functions:

a) Further basic output primitives such as DISJOINT POLYLINE and POLYGON SET.

b) Higher level output primitives such as CIRCLE, RECTANGLE, and ELLIPSE.

c) Extended capabilities in the area of text processing, such as named fonts, changing character set, appended text, and restricted text.

d) Attribute control such as filled-area edge properties, auxiliary colour, and direct colour selection.

e) Enhanced facilities for tailoring and controlling the interpretation of the metafile such as scaling mode, control of the precision of various items, and the control of default values.

This additional functionality of CGM causes no special problems when a CGM is being output from GKS, since such elements would simply not be generated. Thus the interpretation by GKS of a CGM generated by GKS is well defined. The interpretation by GKS of a CGM generated outside of GKS is less well defined. This aspect of the relationship between the two standards is beyond the scope of this annex.

E.4 Mapping Concepts

The tables in later in this annex present mappings between GKS functions and CGM elements. This sub-clause describes the concepts used to derive the mappings.

E.4.1 Principles

The following principles are the basis of the GKS/CGM model and of the function mappings themselves:

a) conceptual compatibility with GKS;

b) compatibility with the design concepts of CGM;

c) compatibility with the Computer Graphics Interface (CGI), of whose functions the CGM elements form a subset;

d) extensibility of the CGM to a GKSM-like metafile.

E.4.2 Workstations

The CGM is generated, in this model, by a workstation of type MO. The behaviour of the workstation, particularly in response to dynamic GKS functions, can be illustrated by analogy: in most respects, the MO/CGM workstation in a level-0 GKS may be implemented in a manner analagous to a workstation of category OUTPUT (e.g., a plotter), whose device instruction set corresponds to the CGM elements. Strategies for correctly sending device instructions to such a real device are similar to those generating the proper elements on the metafile. The CGM is read by a workstation of category MI. Certain elements, such as the metafile descriptor and precision-setting elements, are viewed as directives to the MI workstation itself, so that it may correctly read the metafile contents.

E.4.3 Picture Generation

A metafile is composed of a collection of mutually independent pictures. GKS does not have the concept of "picture" as defined in CGM, but it does formalize the notion of an empty view surface. GKS actions which cause clearing of the view surface, such as CLEAR WORKSTATION, are defined to delimit metafile pictures.

There is another mechanism which leads to generation of pictures in this model of the GKS/CGM relationship. GKS contains functions which have potential dynamic effects on a non-empty view surface. The CGM design concept excludes dynamic modification of pictures. At level 0 of GKS the effects of dynamic functions on the view surface are workstation dependent, and it is not possible for an application to determine the effects. The effects can range from clearing of the view surface to dynamic modification of the image. The former effect is defined for the MO/CGM workstation in this model — a new picture is begun in response to any GKS function that implies dynamic change to a non-empty view surface.

E.4.4 Coordinates and Clipping

The coordinate space of the metafile, VDC, is defined as being identical to the NDC space of GKS. NDC are real numbers in the unit interval. Accordingly, VDC TYPE is set to real in the metafile (the default is integer).

The model of CGM clipping in the GKS environment is similar to the relationship defined in annex E of GKS. Clipping is always 'on' in the metafile, which is the default value of the CLIP INDICATOR element (hence CLIP INDICATOR elements need never be written to the metafile). The CGM CLIP REC-TANGLE element has either the value of the 'clipping rectangle' entry of the GKS state list, or the unit square in VDC, depending upon whether the 'clipping indicator' entry in the GKS state list is 'clip' or 'noclip', respectively. Because the VDC EXTENT element always has the value of the GKS workstation window, the interpreter of the metafile has complete information to achieve GKS clipping.

E.4.5 Workstation Transformation

The workstation transformation is defined in GKS by setting a workstation window in device-independent NDC and a workstation viewport in device-dependent DC. The workstation window is written to the metafile with the VDC EXTENT element. CGM does not have facilities to define the placement of the image on a physical view surface. Hence the workstation viewport of GKS is discarded.

E.4.6 Colour Table

The CGM allows COLOUR TABLE elements to appear in the body of a picture. This obviates the need to gather all colour index definitions for a picture together before putting out the first primitive of a picture. The CGM does not, however, explicitly address the dynamic effects of changing the representation of a colour index to which visible primitives are bound. Any implementation which relies on the COLOUR TABLE element to achieve such dynamic effects within a picture is using information outside of this Standard.

E.4.7 Higher Levels

CGM can be used to capture pictures from implementations of GKS above level 0. The required exten-tions to the picture capture model are considerably more complex than the definition of the level-0 model. They are not defined in detail in this annex, but some conceptual guidelines are provided.

In all respects, the level-1 and level-2 extensions to the MO/CGM workstation may be implemented in a manner analogous to implementing those functions on a non-dynamic workstation of category OUT-PUT.

E.4.7.1 Dynamic Effects. As described in *E.*4.2, dynamic changes to an image cause generation of a new metafile picture. As the CGM lacks segmentation, all previously displayed primitives are lost at level 0.

Segmentation is available at level 1. Dynamic changes can be achieved for those primitives in segments if the implementation begins a new picture and rewrites the primitives of all visible defined segments to the metafile. The effect at metafile interpretation is as if an implicit regeneration had occurred. This approach can be used to realize dynamic effects of all of the functions changing segment attributes, as well as the effects of the SET xxx REPRESENTATION functions.

E.4.7.2 Segment Attribute Functions. Because the CGM lacks any segmentation functions, all segment attribute functions need be resolved before generation of the metafile, i.e., before writing segmented primitives to the metafile. For example, the NDC to NDC segment transformation is applied to all segmented primitives before writing them to the metafile.

E.4.7.3 SET xxx REPRESENTATION Functions. The COLOUR TABLE and PATTERN TABLE elements can be used to communicate the representations (but not the dynamics) of their GKS counterparts through the metafile. For all other values of 'xxx', a different approach is necessary. When primitives are bound to a bundle index whose representation has changed, proper pictures can be defined by setting CGM ASFs 'individual' and sending individual attributes from the GKS state list or the bundle representation, according to the GKS ASFs.

E.5 Metafile Generation

Included in the following tables is a particular set of mappings of the GKS level 0 functions onto CGM elements. Such a set of mappings is not unique, but the mappings presented are usable and suitable for guiding implementation of a CGM generator in a GKS environment. The mapping concepts of *E.*4 are assumed. With the exception of VDC TYPE, the element defaults of clause 6 are assumed as well.

E.5.1 Control Functions

GKS Function	CGM Element	Notes
OPEN WORKSTATION	BEGIN METAFILE	(1)
	(Metafile Descriptor)	(2)
	BEGIN PICTURE	(3)
	BEGIN PICTURE BODY	
CLOSE WORKSTATION	END PICTURE	
	END METAFILE	
ACTIVATE WORKSTATION	Attribute settings	(4)
	CLIP RECTANGLE	(5)
	Enable Output to metafile	
DEACTIVATE WORKSTATION	Disable output to metafile	
CLEAR WORKSTATION	No Action	(6)
	or	
	END PICTURE	
	BEGIN PICTURE	
	VDC EXTENT	(7)
	BACKGROUND COLOUR	(8)
	BEGIN PICTURE BODY	
	Attribute settings	(4)
	CLIP RECTANGLE	(5)
UPDATE WORKSTATION	No action	(9)
ESCAPE	ESCAPE	

NOTES

1. The use of the 'identifier' parameter of BEGIN METAFILE is implementation dependent.

2. See *E*.5.5.

3. The use of the 'identifier' parameter of BEGIN PICTURE is implementation dependent.

4. The attribute settings ensure that the metafile attributes in effect when the first graphical primitive element of a picture is encountered match the current GKS attributes.

5. On activate workstation or when a new picture is begun, a CLIP RECTANGLE element is written to the metafile with value (0.,0.,1.,1.) if the 'clipping indicator' entry in the GKS state list is 'noclip', or with the value of the 'clipping rectangle' in the GKS state list if the 'clipping indicator' entry in the GKS state list is 'clip'. The CLIP INDICATOR element of the metafile is always 'on', which is the default.

6. The action on CLEAR WORKSTATION depends on the control flag. If the control flag is CONDITIONALLY action is only taken if the 'Display Surface Empty' is NOTEMPTY on the workstation. If the control flag is ALWAYS then the elements shown are always generated.

7. VDC EXTENT is included and is set to the current value of the GKS WORKSTATION WINDOW, if that value differs from (0.,0.,1.,1.).

8. If GKS colour index 0 has been explicitly set by the GKS function SET COLOUR REPRESENTATION, then the BACKGROUND COLOUR element should be included (it may be omitted only if the representation matches the CGM default at this point).

9. UPDATE has no graphical effect and has no effect on metafile contents. The MO workstation may, however, synchronize the metafile contents with the application state by flushing any buffered output at this point.

E.5.2 Output Functions

GKS Function	CGM Element	Notes
POLYLINE	POLYLINE	
POLYMARKER	POLYMARKER	
TEXT	TEXT	(1)
FILL AREA	POLYGON	
CELL ARRAY	CELL ARRAY	
GDP	GDP	(2)

NOTES

1. The TEXT flag is set to 'final'.

2. Where there are GDPs registered which correspond to CGM output primitives, the CGM primitives should be used rather than GDPs.

E.5.3 Attributes

GKS Function	CGM Element	Notes
SET POLYLINE INDEX	LINE BUNDLE INDEX	

SET LINETYPE	LINE TYPE	
SET LINEWIDTH SCALE FACTOR	LINE WIDTH	
SET POLYLINE COLOUR INDEX	LINE COLOUR	
SET POLYMARKER INDEX	MARKER BUNDLE INDEX	
SET MARKER TYPE	MARKER TYPE	
SET MARKER SIZE SCALE FACTOR	MARKER SIZE	
SET POLYMARKER COLOUR INDEX	MARKER COLOUR	
SET TEXT INDEX	TEXT BUNDLE INDEX	
SET TEXT FONT AND PRECISION	CHARACTER SET INDEX	(1)
	ALTERNATE CHARACTER SET INDEX	
	TEXT FONT INDEX	
	TEXT PRECISION	
SET CHARACTER EXPANSION FACTOR	CHARACTER EXPANSION FACTOR	
SET CHARACTER SPACING	CHARACTER SPACING	
SET TEXT COLOUR INDEX	TEXT COLOUR	
SET CHARACTER HEIGHT	CHARACTER HEIGHT	
SET CHARACTER UP VECTOR	CHARACTER ORIENTATION	
SET TEXT PATH	TEXT PATH	
SET TEXT ALIGNMENT	TEXT ALIGNMENT	
SET FILL AREA INDEX	FILL BUNDLE INDEX	
SET FILL AREA INTERIOR STYLE	INTERIOR STYLE	
SET FILL AREA STYLE INDEX	HATCH INDEX	(2)
	PATTERN INDEX	(2)
SET FILL AREA COLOUR INDEX	FILL COLOUR	
SET PATTERN SIZE	PATTERN SIZE	
SET PATTERN REFERENCE POINT	FILL REFERENCE POINT	
SET ASPECT SOURCE FLAGS	ASPECT SOURCE FLAGS	
SET COLOUR REPRESENTATION	COLOUR TABLE	

NOTES

1. GKS includes the notion of character set within 'font', whereas CGM separates the two concepts. When the value of 'font' in the GKS state list changes, then the CGM elements FONT INDEX, CHARACTER SET INDEX, and ALTERNATE CHARACTER SET INDEX are written to the metafile, each with the value of the 'font' entry in the GKS state list.

2. Legal values of the GKS 'fill area style index' differ depending upon whether the current interior style is 'hatch' or 'pattern'. Therefore a negative GKS style index results only in the generation of the HATCH INDEX element, and a positive value results in the generation of both the HATCH INDEX and PATTERN INDEX elements.

E.5.4 Transformation Functions

GKS Function	CGM Element	Notes
SET WINDOW (of currently selected normalization transformation)	CHARACTER HEIGHT CHARACTER ORIENTATION PATTERN SIZE FILL REFERENCE POINT	
SET VIEWPORT (of currently selected normalization transformation)	CHARACTER HEIGHT CHARACTER ORIENTATION PATTERN SIZE FILL REFERENCE POINT CLIP RECTANGLE	(1)
SELECT NORMALIZATION TRANSFORMATION	CHARACTER HEIGHT CHARACTER ORIENTATION PATTERN SIZE FILL REFERENCE POINT CLIP RECTANGLE	(1)
SET CLIPPING INDICATOR	CLIP RECTANGLE	(2)
SET WORKSTATION WINDOW	END PICTURE BEGIN PICTURE VDC EXTENT BACKGROUND COLOUR BEGIN PICTURE BODY Attribute Settings CLIP RECTANGLE	(3) (4)
SET WORKSTATION VIEWPORT	As WORKSTATION WINDOW	(3)

NOTES

1. If the 'clipping rectangle' entry in the GKS state list is changed, then a CLIP RECTANGLE element is written to the metafile. The element is written with the value (0.,0.,1.,1.) if the 'clipping indicator' entry in the GKS state list is 'noclip', or with the value of the 'clipping rectangle' in the GKS state list if the 'clipping indicator' entry in the GKS state list is 'clip'.

2. If the 'clipping indicator' entry in the GKS state list is changed, then a CLIP RECTANGLE element is written to the metafile. The element is written with the value (0.,0.,1.,1.) if the 'clipping indicator' entry in the GKS state list is changed to 'noclip', or with the value of the 'clipping rectangle' in the GKS state list if the 'clipping indicator' entry in the GKS state list is changed to 'clip'.

3. The specified actions need only be performed if the view surface is non-empty and the GKS functions imply a dynamic change to the image.

4. On activate workstation or when a new picture is begun, a CLIP RECTANGLE element is written to the metafile with value (0.,0.,1.,1.) if the 'clipping indicator' entry in the GKS state list is 'noclip', or with the value of the 'clipping rectangle' entry in the GKS state list if the 'clipping indicator' entry in the the GKS state list is 'clip'.

E.5.5 Metafile Description

At the head of a metafile is a set of Metafile Descriptor (MD) elements. It is useful to view these elements as forming a Metafile Description Table (similar to the GKS and Workstation Description Tables in GKS).

In the GKS context, the following description table would be written at the beginning of a metafile. For the elements which are listed as "i.d.", it is implementation dependent both whether the elements are included in the table and what values are assigned to the elements.

METAFILE ELEMENT LIST	Elements listed in $E.5.1$-$E.5.5$, or some known subset
METAFILE VERSION	1
VDC TYPE	real
METAFILE DEFAULTS REPLACEMENT	i.d.
METAFILE DESCRIPTION	i.d.
INTEGER PRECISION	i.d.
REAL PRECISION	i.d.
INDEX PRECISION	i.d.
COLOUR PRECISION	i.d.
COLOUR INDEX PRECISION	i.d.
MAXIMUM COLOR INDEX	i.d.
COLOUR VALUE EXTENT	i.d.
FONT LIST	i.d.
CHARACTER CODING ANNOUNCER	i.d.
CHARACTER SET LIST	i.d.

i.d. = implementation dependent

The Metafile Descriptor elements are discussed in sub-clause 5.3. METAFILE VERSION and METAFILE ELEMENT LIST are mandatory. As indicated in the discussion of coordinate systems in $E.4$, VDC TYPE is included in order to declare real coordinates. With the exception of VDC TYPE, all metafile defaults are satisfactory. Inclusion of the METAFILE DEFAULTS REPLACEMENT element to change any control, picture descriptor, and attribute defaults is optional and implementation dependent. It is also implementation dependent whether the CGM generator includes any of the other MD elements, such as the precision-setting elements.

E.6 Interpretation of CGM by GKS

This sub-clause complements the preceding one by describing how the metafile elements, generated by a GKS program according to the mappings described, are subsequently interpreted by the GKS INTERPRET ITEM function and/or the MI workstation.

A number of the elements below are specified as causing GKS state list entries to be set, and have parameters specified in VDC (which is chosen to be identical to GKS NDC). The GKS state list entries are in WC. The VDC (NDC) are mapped by the inverse of the current normalization transformation before the GKS state list values are set.

BEGIN METAFILE
    The first item interpreted. The metafile description table immediately follows. Its elements inform the MI workstation how to read the metafile.

END METAFILE
    No further items may be read.

BEGIN PICTURE
    Appropriate GKS state list values are set to correspond to CGM defaults. Appropriate workstation state list values on active OUTPUT and OUTIN workstations are set to correspond to CGM defaults. It is not intended that this action, or the interpretation of any picture descriptor elements, cause any immediate dynamic changes to the view surface, which is cleared upon BEGIN PICTURE BODY — the implementation may wish to buffer these actions to suppress such changes, if such changes are undesirable. Only picture descriptor elements may be interpreted until BEGIN PICTURE BODY.

BEGIN PICTURE BODY
    Causes a CLEAR WORKSTATION on all active workstations.

VDC EXTENT
    Causes the workstation window on all active workstations to be set to the rectangle specified by the element.

BACKGROUND COLOUR
    Resets colour index 0 appropriately on all active workstations.

CLIP RECTANGLE
    Sets the 'clipping rectangle' entry in the GKS state list to the rectangle specified by the metafile element.

POLYLINE
    Generates a GKS polyline primitive.

POLYMARKER
    Generates a GKS polymarker primitive.

POLYGON
    Generates a GKS fill area primitive.

TEXT
    Generates a GKS text primitive.

CELL ARRAY
    Generates a GKS cell array primitive.

GDP
    Generates a GKS GDP.

Further output primitives in CGM (e.g., CIRCLE)
    Mapped into and generates corresponding GKS GDP, if it is known to the implementation.

ASPECT SOURCE FLAGS
LINE BUNDLE INDEX
LINE TYPE
LINE WIDTH
LINE COLOUR

MARKER BUNDLE INDEX
MARKER TYPE
MARKER SIZE
MARKER COLOUR
FILL BUNDLE INDEX
INTERIOR STYLE
FILL COLOUR
TEXT BUNDLE INDEX
CHARACTER EXPANSION FACTOR
CHARACTER SPACING
TEXT COLOUR
TEXT PATH
TEXT ALIGNMENT
> The preceding attribute elements have a clear one-to-one mapping with GKS attributes. They cause the corresponding entries to be set in the GKS state list.

HATCH INDEX
PATTERN INDEX
> Both elements cause the 'fill area style index' entry to be set appropriately in the GKS state list.

FILL REFERENCE POINT
> Sets the 'pattern reference point' entry in the GKS state list.

PATTERN SIZE
> Sets the 'pattern size vectors' entry in the GKS state list.

CHARACTER SET INDEX
ALTERNATE CHARACTER SET INDEX
> May affect 'text font and precision' entry in the GKS state list.

TEXT FONT INDEX
TEXT PRECISION
> Set 'text font and precision' entry in the GKS state list.

CHARACTER HEIGHT
> Sets the 'character height' and computes and sets the 'character width' entries in the GKS state list.

CHARACTER ORIENTATION
> Sets the 'character height vector' and 'character width vector' entries in the GKS state list.

COLOUR TABLE
> Sets appropriate entries in the table of colour representations in the workstation state lists of all active workstations.

ESCAPE
> Causes the appropriate ESCAPE function to be generated.

ANSI X3.122 - 1986

Information Processing Systems

Computer Graphics

Metafile for the Storage and Transfer
of Picture Description Information

Part 2

Character Encoding

ANSI X3.122 - 1986
Part 2

CONTENTS

0 Introduction

0.1 Purpose of the Character Encoding

The Character Encoding of the Computer Graphics Metafile (CGM) provides a representation of the Metafile syntax intended for situations in which it is important to minimize the size of the metafile or transmit the metafile through character-oriented communications services. The encoding uses compact representation of data that is optimized for storage or transfer between computer systems.

If minimizing the processing overhead is more important than data compaction, an encoding such as the Binary Encoding contained in part 3 of this Standard may be more appropriate. If human readability is the most important criterion, an encoding such as the Clear Text Encoding in part 4 of this Standard may be more appropriate.

0.2 Objectives

This encoding was designed with the following objectives:

a) regular syntax: All elements of the metafile should be encoded in a uniform way so that parsing the metafile is simple;

b) compactness: The encoding should provide a highly compact metafile, suitable for systems with restricted storage capacity or transfer bandwidth;

c) extensibility: the encoding should allow for future extensions;

d) transportability: the encoding should be suitable for use with transport mechanisms designed for character-oriented data based on a standard national character set derived from ISO 646.

0.3 Metafile Characteristics

Each CGM command follows a simple regular syntax. Thus, new commands can be added in a future revision of this Standard such that existing CGM interpreters can recognize (and ignore) the new commands. Also, new operands can be added to existing commands in the future revision of the standard such that existing CGM interpreters can recognize (and ignore) the additional operands.

Each CGM operand follows a simple regular syntax. Operands are variable in length. This permits small values to be represented by the smallest number of bytes.

A certain range of operand values of standard commands have been reserved for private use; the remaining range is either standardized or reserved for future standardization.

0.4 Relationship to Other Standards

The Character Encoding has been developed in collaboration with the ISO subcommittee responsible for character sets and coding (ISO TC97/SC2/WG8), ECMA, and CEPT. The encoding conforms to the rules for code extension specified in ISO 2022 in the category of complete coding system.

The representation of character data in this part of the Standard follows the rules of ISO 646 and ISO 2022.

For certain elements, the CGM defines value ranges as being reserved for registration. The values and their meanings will be defined using the established procedures (see part 1, 4.11.)

0.5 Status of Annexes

The annexes do not form an integral part of the Standard but are included to provide extra information and explanation.

1 Scope and Field of Application

1.1 Scope

This part of this Standard specifies a character encoding of the Computer Graphics Metafile. For each of the elements specified in part 1 of this Standard, an encoding is specified.

1.2 Field of Application

This encoding of the Computer Graphics Metafile provides a highly compact representation of the metafile, suitable for applications that require the metafile to be of minimum size and suitable for transmission with character-oriented transmission services.

2 References

ISO 646 Information Processing - 7-bit coded character set for information interchange

ISO 2022 Information Processing - ISO 7-bit and 8-bit coded character sets - Code extension techniques

ISO 2375 Information Processing - Character set registration

ISO 6429 Information Processing - ISO 7-bit and 8-bit coded character sets - Additional control functions for character-imaging devices

ECMA 96 Graphics Data Syntax for a multiple Workstation Interface

CEPT Rev. of T/CD 6.1 Videotex Presentation Layer Data Syntax

3 Notational Conventions

3.1 7-Bit and 8-Bit Code Tables

The bits of the bit combinations of the 7-bit code are identified by b7, b6, b5, b4, b3, b2, and b1, where b7 is the highest-order, or most-significant, bit and b1 is the lowest-order, or least-significant, bit.

The bit combinations may be interpreted to represent integers in the range 0 to 127 in binary notation by attributing the following weights to the individual bits:

Bit:	b7	b6	b5	b4	b3	b2	b1
Weight:	64	32	16	8	4	2	1

In this Standard, the bit combinations are identified by notation of the form x/y, where x is a number in the range 0 to 7 and y is a number in the range 0 to 15. The correspondence between the notations of the form x/y and the bit combinations consisting of the bits b7 to b1 is as follows:

— x is the number represented by b7, b6, and b5 where these bits are given the weights 4, 2, and 1 respectively;

— y is the number represented by b4, b3, b2, and b1 where these bits are given the weights 8, 4, 2, and 1 respectively.

The notations of the form x/y are the same as those used to identify code table positions, where x is the column number and y is the row number.

A 7-bit code table consists of 128 positions arranged in eight columns and sixteen rows. The columns are numbered 0 to 7 and the rows are numbered 0 to 15. Figure 1 shows a 7-bit code table.

An example illustrates the 7-bit code: "1/11" refers to the bit combination in column 1, row 11 of the code table, binary 0011011.

The bits of the bit combinations of the 8-bit code are identified by b8, b7, b6, b5, b4, b3, b2, and b1, where b8 is the highest-order, or most-significant, bit and b1 is the lowest-order, or least-significant, bit.

The bit combinations may be interpreted to represent integers in the range 0 to 255 in binary notation by attributing the following weights to the individual bits:

Bit:	b8	b7	b6	b5	b4	b3	b2	b1
Weight:	128	64	32	16	8	4	2	1

Using these weights, the bit combinations of the 8-bit code are interpreted to represent numbers in the range 0 to 255.

In this Standard, the bit combinations are identified by notation of the form xx/yy, where xx and yy are numbers in the range 00 to 15. The correspondence between the notations of the form xx/yy and the bit combinations consisting of the bits b8 to b1 is as follows:

— xx is the number represented by b8, b7, b6, and b5 where these bits are given the weights 8, 4, 2, and 1 respectively;

— yy is the number represented by b4, b3, b2, and b1 where these bits are given the weights 8, 4, 2, and 1 respectively.

The notations of the form xx/yy are the same as those used to identify code table positions, where xx is the column number and yy is the row number. An 8-bit code table consists of 256 positions arranged in sixteen columns and sixteen rows. The columns and rows are numbered 00 to 15. Figure 2 shows an 8-bit code table.

An example illustrates the 8-bit code: 04/01 represents the 8-bit byte 01000001, whereas 4/1 represents the 7-bit byte 1000001.

3.2 Code Extension Techniques Vocabulary

In describing the characters that may occur within string parameters, certain terms imported from other standards (e.g., ISO 2022) are useful. In the context of the CGM, these terms, and the concepts to which they refer, apply only within the string parameters of the TEXT, APPEND TEXT, and RESTRICTED TEXT metafile elements.

3.2.1 C0 Sets

A C0 set is a set of 30 control characters represented in a 7-bit code by 0/0 to 1/15, except 0/14 and 0/15 which shall be unused, and in an 8-bit code by 00/00 to 01/15, except 00/14 and 00/15 which shall be usused. C0 sets occupy columns 0 and 1 of a 7-bit code table or columns 00 and 01 of an 8-bit code table. The meanings of C0 controls within string parameters are described in 6.9.3.

3.2.2 C1 Sets

A C1 set is a set of up to 32 control characters represented by bit combinations 08/00 to 09/15 in an 8-bit code. C1 sets occupy columns 08 and 09 of the 8-bit code table. In a 7-bit code the C1 control functions are represented by 2-byte escape sequences. This CGM encoding reserves the bit combinations 9/8 and 9/12 (ESC 5/8 and ESC 5/12 in a 7-bit environment, ESC = 1/11); these shall not be part of the content of string parameters. Other C1 control characters from other standards, such as ISO 6429, may be used within string parameters by agreement between the interchanging parties.

3.2.3 G-Sets

The G-sets (G0, G1, G2, G3) are coded character sets of 94 or 96 characters. CHARACTER SET INDEX designates which character set is to be the G0 set. ALTERNATE CHARACTER SET INDEX designates a character set to be used as both the G1 and G2 sets. The G-sets may be "invoked into" (caused to occupy) columns 2 through 7 of a 7-bit code table, or columns 02 through 07 and 10 through 15 of an 8-bit code table. This encoding of the CGM uses the G0 and G1/G2 sets within string parameters. The G3 set may be used within the string parameters of conforming metafiles; this requires selection of the extended 7-bit or extended 8-bit mode in the CHARACTER CODING ANNOUNCER. The CGM does not provide an element to explicitly designate the G3 sets; this may be done within a text string in accordance with ISO 2022, or by other means agreed upon by the interchanging parties.

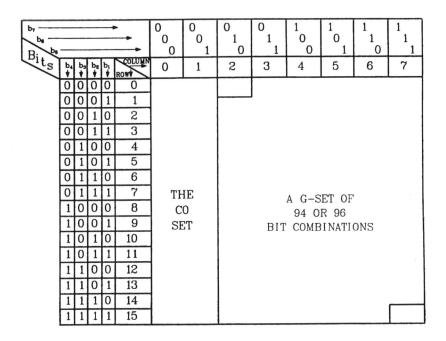

Figure 1. The 7-bit code table.

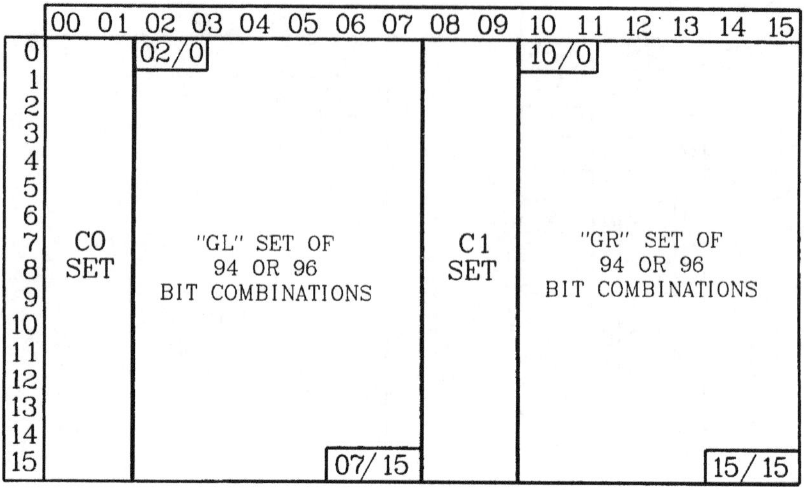

	00 01	02 03 04 05 06 07	08 09	10 11 12 13 14 15
0		02/0		10/0
1				
2				
3				
4				
5				
6				
7	CO	"GL" SET OF	C1	"GR" SET OF
8	SET	94 OR 96	SET	94 OR 96
9		BIT COMBINATIONS		BIT COMBINATIONS
10				
11				
12				
13				
14				
15		07/15		15/15

Figure 2. The 8-bit code table.

4 Entering and Leaving the Metafile Environment

4.1 Implicitly Entering the Metafile Environment

The CGM coding environment may be entered implicitly, by agreement between the interchanging parties. This is suitable only if there is not to be any interchange with services using other coding techniques.

4.2 Designating and Invoking the CGM Coding Environment From ISO 2022

For interchange with services using the code extension techniques of ISO 2022, the CGM coding environment shall be designated and invoked from ISO 2022 environment by the following escape sequence:

ESC 2/5 F

where ESC is the bit combination 1/11, and F refers to a bit combination that will be assigned by the ISO Registration Authority for ISO 2375.

The first bit combination occuring after this escape sequence will then represent the opcode of a CGM metafile element.

After the end of one or more metafiles (i.e., after the END METAFILE element) or between pictures (i.e., after the END PICTURE element), the following escape sequence may be used to return to the ISO 2022 coding environment:

ESC 2/5 4/0

This not only returns to the ISO 2022 coding environment, but also restores the designation and invocation of coded character sets to the state that existed prior to entering the CGM coding environment with the ESC 2/5 F sequence. (The terms "designation" and "invocation" are defined in ISO 2022.)

5 Method of Encoding Opcodes

Each metafile element is composed of one opcode and parameters as required. The opcodes are coded as a sequence of bit combinations from columns 2 and 3 of the code chart. The encoding technique supplies:

— the basic opcode set;

— extension opcode sets.

5.1 Encoding Technique of the Basic Opcode Set

The basic opcode set consists of single-byte and double-byte opcodes. Single-byte opcodes are from column 2 of the code chart. Bits b4 to b1 are used to encode the opcode. The format is as follows:

```
b8              b1
+-+-----+-------+
|X|0 1 0|b b b b|
+-+-----+-------+
```

The "X" bit (bit b8) is the parity bit (or omitted bit) in a 7-bit environment. In an 8-bit environment it is 0. For double-byte opcodes the first byte is from column 3 and the second byte is from column 2 or 3 of the code chart. Bits b4 to b1 of the first byte and bits b5 to b1 of the second byte are used to encode the opcode:

```
b8              b1   b8              b1
+-+-----+-------+    +-+---+---------+
|X|0 1 1|b b b b|    |X|0 1|b b b b b|
+-+-----+-------+    +-+---+---------+
```

The bit combination 3/15, the EXTEND OPCODE SPACE (EOS) allows extension of the basic opcode space (see 5.2).

The basic opcode set, supplied by this encoding technique consists of 496 opcodes, being:

— 16 single-byte opcodes (from column 2);

— 15 x 32 = 480 double-byte opcodes (first byte from column 3 except bit combination 3/15, second byte from column 2 or 3).

5.2 Extension Mechanism

The basic opcode set can be extended with an unlimited number of extension opcode sets by means of the EXTEND OPCODE SPACE code (EOS, 3/15).

The N-th extension opcode set consists of opcodes of the basic opcode set, prefixed with n times the code EOS. The three possible formats of an opcode from the N-th extension opcode set are:

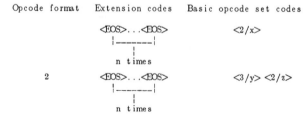

```
Opcode format    Extension codes    Basic opcode set codes

                 <EOS>...<EOS>           <2/x>
                 |---------|
                      |
                   n times

      2          <EOS>...<EOS>       <3/y> <2/z>
                 |---------|
                      |
                   n times
```

<EOS>...<EOS>   <3/y> <3/z>
|-------|
|
n times

<EOS> = 3/15
x   = 0,1,..,15
y   = 0,1,..,14
z   = 0,1,..,15
n   = 0,1,.....

n = 0 selects the basic opcode set.
n = 1 selects the first extension opcode set.
n = N selects the N-th extension opcode set.

The number of opcodes supplied by this encoding technique (basic opcode set plus extension opcode sets) is 496*(n+1), where n is the number of extension sets. (Each extension set has 496 opcodes — 16 single-byte opcodes plus 480 double-byte codes.)

5.3 Opcode Assignments

Table 1 lists the opcode assignments for the CGM elements. All opcodes are from the basic opcode set. They have be organized as follows: single-byte opcodes are assigned to Graphical Primitive elements except for some of the Circular and Elliptical output elements, where double-byte opcodes with combination 3/4 for the first byte have been assigned. All other metafile elements have double-byte opcodes with the following bit combinations for the first byte:

3/0	for Delimiter Elements
3/1	for Metafile Descriptor Elements
3/2	for Picture Descriptor Elements
3/3	for Control Elements
3/5 and 3/6	for Attribute Elements
3/7	for Escape and External Elements

TABLE 1. Opcodes for Metafile Elements.				
Opcode	7-Bit Coding		8-Bit Coding	
BEGIN METAFILE opcode	3/0	2/0	03/0	02/0
END METAFILE opcode	3/0	2/1	03/0	02/1
BEGIN PICTURE opcode	3/0	2/2	03/0	02/2
BEGIN PICTURE BODY opcode	3/0	2/3	03/0	02/3
END PICTURE opcode	3/0	2/4	03/0	02/4
METAFILE VERSION opcode	3/1	2/0	03/1	02/0
METAFILE DESCRIPTION opcode	3/1	2/1	03/1	02/1
VDC TYPE opcode	3/1	2/2	03/1	02/2
INTEGER PRECISION opcode	3/1	2/3	03/1	02/3
REAL PRECISION opcode	3/1	2/4	03/1	02/4
INDEX PRECISION opcode	3/1	2/5	03/1	02/5
COLOUR PRECISION opcode	3/1	2/6	03/1	02/6
COLOUR INDEX PRECISION opcode	3/1	2/7	03/1	02/7
MAXIMUM COLOUR INDEX opcode	3/1	2/8	03/1	02/8
COLOUR VALUE EXTENT opcode	3/1	2/9	03/1	02/9
METAFILE ELEMENT LIST opcode	3/1	2/10	03/1	02/10
BEGIN METAFILE DEFAULTS REPLACEMENT opcode	3/1	2/11	03/1	02/11
END METAFILE DEFAULTS REPLACEMENT opcode	3/1	2/12	03/1	02/12
FONT LIST opcode	3/1	2/13	03/1	02/13
CHARACTER SET LIST opcode	3/1	2/14	03/1	02/14
CHARACTER CODING ANNOUNCER opcode	3/1	2/15	03/1	02/15
SCALING MODE opcode	3/2	2/0	03/2	02/0
COLOUR SELECTION MODE opcode	3/2	2/1	03/2	02/1
LINE WIDTH SPECIFICATION MODE opcode·	3/2	2/2	03/2	02/2
MARKER SIZE SPECIFICATION MODE opcode	3/2	2/3	03/2	02/3
EDGE WIDTH SPECIFICATION MODE opcode	3/2	2/4	03/2	02/4
VDC EXTENT opcode	3/2	2/5	03/2	02/5
BACKGROUND COLOUR opcode	3/2	2/6	03/2	02/6
VDC INTEGER PRECISION opcode	3/3	2/0	03/3	02/0
VDC REAL PRECISION opcode	3/3	2/1	03/3	02/1
AUXILIARY COLOUR opcode	3/3	2/2	03/3	02/2
TRANSPARENCY opcode	3/3	2/3	03/3	02/3
CLIP RECTANGLE opcode	3/3	2/4	03/3	02/4
CLIP INDICATOR opcode	3/3	2/5	03/3	02/5
POLYLINE opcode	2/0		02/0	
DISJOINT POLYLINE opcode	2/1		02/1	
POLYMARKER opcode	2/2		02/2	
TEXT opcode	2/3		02/3	
RESTRICTED TEXT opcode	2/4		02/4	
APPEND TEXT opcode	2/5		02/5	
POLYGON opcode	2/6		02/6	
POLYGON SET opcode	2/7		02/7	
CELL ARRAY opcode	2/8		02/8	
GENERALIZED DRAWING PRIMITIVE opcode	2/9		02/9	
RECTANGLE opcode	2/10		02/10	
CIRCLE opcode	3/4	2/0	03/4	02/0
CIRCULAR ARC 3 POINT opcode	3/4	2/1	03/4	02/1
CIRCULAR ARC 3 POINT CLOSE opcode	3/4	2/2	03/4	02/2
CIRCULAR ARC CENTRE opcode	3/4	2/3	03/4	02/3

CIRCULAR ARC CENTRE CLOSE opcode	3/4	2/4	03/4	02/4
ELLIPSE opcode	3/4	2/5	03/4	02/5
ELLIPTICAL ARC opcode	3/4	2/6	03/4	02/6
ELLIPTICAL ARC CLOSE opcode	3/4	2/7	03/4	02/7
LINE BUNDLE INDEX opcode	3/5	2/0	03/5	02/0
LINE TYPE opcode	3/5	2/1	03/5	02/1
LINE WIDTH	3/5	2/2	03/5	02/2
LINE COLOUR opcode	3/5	2/3	03/5	02/3
MARKER BUNDLE INDEX opcode	3/5	2/4	03/5	02/4
MARKER TYPE opcode	3/5	2/5	03/5	02/5
MARKER SIZE opcode	3/5	2/6	03/5	02/6
MARKER COLOUR opcode	3/5	2/7	03/5	02/7
TEXT BUNDLE INDEX opcode	3/5	3/0	03/5	03/0
TEXT FONT INDEX opcode	3/5	3/1	03/5	03/1
TEXT PRECISION opcode	3/5	3/2	03/5	03/2
CHARACTER EXPANSION FACTOR opcode	3/5	3/3	03/5	03/3
CHARACTER SPACING opcode	3/5	3/4	03/5	03/4
TEXT COLOUR opcode	3/5	3/5	03/5	03/5
CHARACTER HEIGHT opcode	3/5	3/6	03/5	03/6
CHARACTER ORIENTATION opcode	3/5	3/7	03/5	03/7
TEXT PATH opcode	3/5	3/8	03/5	03/8
TEXT ALIGNMENT opcode	3/5	3/9	03/5	03/9
CHARACTER SET INDEX opcode	3/5	3/10	03/5	03/10
ALTERNATE CHARACTER SET INDEX opcode	3/5	3/11	03/5	03/11
FILL BUNDLE INDEX opcode	3/6	2/0	03/6	02/0
INTERIOR STYLE opcode	3/6	2/1	03/6	02/1
FILL COLOUR opcode	3/6	2/2	03/6	02/2
HATCH INDEX opcode	3/6	2/3	03/6	02/3
PATTERN INDEX opcode	3/6	2/4	03/6	02/4
EDGE BUNDLE INDEX opcode	3/6	2/5	03/6	02/5
EDGE TYPE opcode	3/6	2/6	03/6	02/6
EDGE WIDTH opcode	3/6	2/7	03/6	02/7
EDGE COLOUR opcode	3/6	2/8	03/6	02/8
EDGE VISIBILITY opcode	3/6	2/9	03/6	02/9
FILL REFERENCE POINT opcode	3/6	2/10	03/6	02/10
PATTERN TABLE opcode	3/6	2/11	03/6	02/11
PATTERN SIZE opcode	3/6	2/12	03/6	02/12
COLOUR TABLE opcode	3/6	3/0	03/6	03/0
ASPECT SOURCE FLAGS opcode	3/6	3/1	03/6	03/1
ESCAPE opcode	3/7	2/0	03/7	02/0
DOMAIN RING opcode	3/7	3/0	03/7	03/0
MESSAGE opcode	3/7	2/1	03/7	02/1
APPLICATION DATA opcode	3/7	2/2	03/7	02/2

6 Method of Encoding Parameters

The parameter part of a CGM element may contain one or more parameters, each parameter consisting of one or more bytes.

All parameters are coded in columns 4 through 7. (However, the coded representation of a 'string' parameter may include bit combinations from other columns of the code table - see the description of string parameters in 6.9). The general format of a parameter byte is:

```
b8                b1
+-+-+-----------+
|X|1|b b b b b b|
+-+-+-----------+
```

The "X" (bit b8) is the parity bit (or omitted bit) in a 7-bit environment. In an 8-bit environment it is 0. Bit b7 is the parameter flag.

Except for 'string' and 'data record' parameters all parameters are encoded using one or both of two formats, Basic format or Bitstream format.

6.1 Basic Format

Each Basic format parameter is coded as a sequence of one or more bytes, structured as follows:

```
b8            b1
+-+-+-+-+-------+
|X|1|e|s|b b b b|   first byte
+-+-+-+-+-------+
```

```
b8            b1
+-+-+-+---------+
|X|1|e|b b b b b|   last byte
+-+-+-+---------+
```

The "X" (bit b8) is the parity bit (or omitted bit) in a 7-bit environment. In an 8-bit environment it is 0. Bit b7 is the parameter flag.

"e" (b6 of each byte) is the extension flag. For single byte parameters, the extension flag is 0. In multi-byte parameters, the extension flag is 1 in all bytes except the last byte, where it is 0.

Bits b5 through b1 are the data bits of the parameter.

"s" is the sign bit; if equal to 0 then the integer is non-negative and if equal to 1 then the integer is negative. The number zero shall always be coded as "plus zero" 4/0. The "minus zero" coding is reserved for special usage (see 6.6.3).

The Basic format is used to encode:

a) enumerated types (E);

b) colour indices (CI);

c) indices other than colour indices (IX);

d) integers (I);

e) real numbers (R);

f) non-incremental coordinates.

The most significant part of the parameter is coded in the first byte. The least significant part of the parameter is coded in the last byte.

6.2 Bitstream Format

Each Bitstream format parameter is encoded as a sequence of one or more bytes, structured as follows:

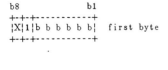

```
b8              b1
+-+-+-----------+
|X|1|b b b b b b|   first byte
+-+-+-----------+
```

```
b8              b1
+-+-+-----------+
|X|1|b b b b b b|   last byte
+-+-+-----------+
```

The "X" is the parity bit (or omitted bit) in a 7-bit environment. In an 8-bit environment it is 0. Bit b7 is the parameter flag.

Bits b6 through b1 are the data bits of the parameter.

The Bitstream format is used to encode:

a)   incremental mode coordinates (see 6.6.2);

b)   colour direct (see 6.8);

c)   colour index lists (see 6.8).

Bitstream data are packed in consecutive databits starting from high-numbered bits to lower-numbered bits of the first byte for the most significant part of the bitstream data.

The end of a Bitstream format parameter cannot be derived from the Bitstream format itself (the format is not self-delimiting). Instead,

—   for incremental mode coordinates, the end of the data (which identifies the end of the Bitstream format parameter) is identified by the <End of Block> code;

—   for colour index lists, the number of bits needed to encode the colour index list (which identifies the end of the Bitstream format parameter) is set by the COLOUR INDEX PRECISION element, or by the LOCAL COLOUR PRECISION parameter (for those elements which contain such a parameter);

—   for colour direct data, the number of bits needed to encode the data (which identifies the end of the Bitstream format operand) is set by the COLOUR PRECISION element, or by the LOCAL COLOUR PRECISION parameter (for those elements which contain such a parameter).

6.3 Coding Integers

Integers are coded as sequences of bytes in the range from 4/0 to 7/15 in the Basic format. If a byte is from columns 4 or 5 of the code table, it is either the last byte in the integer's coded representation or it is a single-byte integer. A multi-byte integer begins with a byte from columns 6 or 7 of the code table.

The structure of integer parameters is as illustrated for Basic format (see 6.1). "bbb..." are bits representing the magnitude of the integer. The most significant part of the parameter is coded in the first byte. The least significant part of the parameter is coded in the last byte.

Any integer can be coded with leading most significant bits which are all zero. For example, 4/3 and 6/0 6/0 4/3 are both valid codings for the integer "+3"; however, efficient metafile generators should avoid such redundant codings.

The size of integer parameters is limited by the current INTEGER PRECISION value.

Integers in the range of -15 to +15 can be coded as single bytes.

(integer: +1) = 4/1   (integer: -1) = 5/1
(integer: +15) = 4/15   (integer: -15) = 5/15

Larger integers require more bytes.

(integer: +16) = 6/0 5/0   (integer: -16) = 7/0 5/0
(integer: +1034) = 6/1 6/0 4/10   (integer: -1034) = 7/1 6/0 4/10

6.4 Coding Real Numbers

Each real number is coded as an integer mantissa followed by an optional exponent, both coded in the Basic format. The exponent is the power of two by which the integer mantissa is to be multiplied.

The exponent may be implicitly defined as a default exponent, which is then omitted in the Real format, or the exponent may be coded explicitly as the second part of the Real format:

<real format> = <mantissa part> [<exponent part>]

Depending on the "exponent allowed" parameter of the REAL PRECISION element, one of the bits in the first byte of the mantissa tells whether the exponent follows. If the "exponent follows" bit in the mantissa is zero, or if REAL PRECISION has specified that there is to be no "exponent follows" bit, then the exponent is omitted and a default value is assumed which is set by another parameter of the REAL PRECISION element.

Note that the coding of real VDC coordinates is controlled by VDC REAL PRECISION, not by REAL PRECISION, and that the rules for default exponents in real VDC coordinates are slightly different (see 6.5 and 6.6).

The mantissa is an integer which is coded in the Basic format. The first byte takes one of two forms, depending on whether or not REAL PRECISION has specified that an "exponent follows" bit is to be included. The format is as follows (first byte):

```
b8              b1
+-+-+-+-+-+-----+   {if explicit exponent allowed = 'allowed',
|X|1|e|s|p|b b b|    i.e. the exponent-follows bit is present}
+-+-+-+-+-+-----+

or

b8              b1
+-+-+-+-+-------+   {if explicit exponent allowed = 'forbidden',
|X|1|e|s|b b b b|    i.e. there is no exponent-follows bit}
+-+-+-+-+-------+


b8              b1
+-+-+-+---------+
|X|1|e|b b b b b|   {last byte}
+-+-+-+---------+
```

"e" is the extension flag, see 6.1 for a description.

"s" is the sign bit (bit b5 of the first byte): 0 for a positive mantissa, 1 for a negative mantissa.

Here "ᴾ" is the exponent-follows bit. If the current "explicit exponent allowed" value of the REAL PRE-CISION element is 'allowed', bit b4 of the first byte of a mantissa is used as the "exponent follows" bit:

1 if an explicit exponent follows, 0 if no exponent follows the mantissa (then the default exponent set by REAL PRECISION is assumed). If the current "explicit exponent allowed" value of the REAL PRECISION element is 'forbidden', bit b4 of the first byte is used as a databit and the default exponent set by REAL PRECISION is assumed.

The exponent is coded as an integer in Basic format, see 6.3.

Mantissas or exponents of "minus zero" are not allowed and reserved for future use.

For example, suppose that REAL PRECISION has specified that each real parameter is to be coded with an "exponent follows" bit in its mantissa. In that case, the binary numeral +1.110010110011 is coded as follows:

```
(real: binary +1.11 00101 10011)

= (mantissa: +111 00101 10011, "exponent-follows")
  (exponent: -12)

  +-+---+-+-+-----+ +-+-+-+---------+ +-+-+-+---------+
= |X|1|1|0|1|1 1 1| |X|1|1|0 0 1 0 1| |X|1|0|1 0 0 1 1| {mantissa}
  +-+---+-+-+-----+ +-+-+-+---------+ +-+-+-+---------+
  +-+-+-+-+-------+
  |X|1|0|1|1 1 0 0| {exponent}
  +-+-+-+-+-------+

= 6/15   6/5   5/3   5/12
```

Again, suppose that REAL PRECISION has specified that real parameters are to be coded without exponents and without exponent-follows bits in their mantissas, and that the default exponent is to be -13. In that case, binary +1.110010110011 is coded as follows:

```
(real: binary +1.110010110011)

= (mantissa: +1110 01011 00110)

  +-+-+-+-+-------+ +-+-+-+---------+ +-+-+-+---------+
= |X|1|1|0|1 1 1 0| |X|1|1|0 1 0 1 1| |X|1|0|0 0 1 1 0| {mantissa}
  +-+-+-+-+-------+ +-+-+-+---------+ +-+-+-+---------+

= 6/14   6/11   4/6
```

6.5 Coding VDCs and Points

A point is a pair of VDC scalars. A VDC scalar is either an integer or real number according to whether VDC TYPE is integer or real.

When VDC TYPE is integer, the encodings of the VDC and point data types are as described in 6.3, Coding Integers. The size of the VDC and point parameters is limited by the current VDC INTEGER PRECISION value.

When VDC TYPE is real, the encodings of the VDC and point data types are as described in 6.4, Coding Real Numbers. The size of the VDC and point parameters is limited by the current VDC REAL PRECISION value. Whether or not "exponent follows" bits are included in the mantissas of VDCs and points is determined by a parameter of VDC REAL PRECISION. The default value for an omitted exponent in a VDC is determined by the "default exponent" parameter in VDC REAL PRECISION. If the exponent is omitted from a point data type, a default exponent is assumed as follows.

— If the point is not in a point list, or is the first point of a point list, the omitted exponent assumes the default value exponents of VDC parameters, as specified by VDC REAL PRECISION.

— If the point is in a point list, but is not the first point of that point list, an omitted exponent may assume a different value as follows: if the exponent is omitted from the x-component of a real point, it defaults to the value of the exponent in the preceding x-component; similarly, an exponent omitted from a y-component assumes the value of the exponent in the preceding y-coordinate.

6.6 Coding Point List Parameters

Point lists may be coded with one or both of the following coding structures. These structures are called 'Displacement Mode' and 'Incremental Mode'. Both formats may be used within a single point list.

Individual points and the first point of each point list are always be coded in Displacement Mode. The displacement values delta-x and delta-y are displacements measured from the origin, i.e. absolute positions. For points after the first point in a point list, each displacement is measured from the preceding point of the point list (this is true regardless of whether the preceding point was coded in Displacement Mode or in Incremental Mode).

6.6.1 Displacement Mode

In Displacement Mode, each point is coded as a sequence of two VDCs. The first VDC value gives the x-component of the point's displacement from the preceding point, while the second VDC value specifies the y-component of that displacement.

<P> = <VDC: delta x> <VDC: delta y>

6.6.2 Incremental Mode

The Incremental Mode is defined as a Differential Chain Code (DCC). The data in this mode does not reflect actual coordinates, but defines steps (increments) from one coordinate position to another. These increments are identified by points on a Ring. A Ring is a set of points on a square whose centre is the previously identified point. The first centre point is encoded in Displacement Mode.

A Ring is characterized by its Radius (R) in Basic Grid Units (BGU), its Angular resolution (by a factor p) and its Direction (D). The maximum number of points on a Ring is 8R. The actual number of points on a Ring with a given Angular resolution factor p follows from:

$$N = \frac{8R}{2^p}, \quad p = 0,1,2,3$$

N is required to be even.

The points on the Ring are numbered, starting at the Direction point D, counter clockwise from 0 to M-1 and clockwise from -1 to -M, with M=N/2.

Following are examples of two Rings, both having Radius R=3. The Ring on the left has Angular resolution factor p=0 and the one on the right has Angular resolution factor p=1.

```
        8   7   6   5   4                    4         2
    9 .   .   .   .   .   .   . ɔ            .         .

   10 .                   . 2      5 .

   11 .

  -12 .           .       . 0     -6 .         .         . 0
              centre                       centre
  -11 .                   . -1

  -10 .                   . -2     -5 .                  . -1

   -9 .   .   .   .   .   .   . -3          .    .    .
        -8  -7  -6  -5  -4                 -4   -3   -2
```

The Direction of the Ring is given by the position of the point with number ZERO. The initial position of this point is on the positive x-axis of VDC space, while the Cartesian axes are drawn through the centre point of the Ring. The Direction of the Rings following the initial one is dependent on the direction of the increments. This Direction is determined in the following way:

Let P2 be the current centre point and P1 the previous one, i.e., P2 is a point on the Ring with centre in P1. The Direction of the Ring is obtained by drawing a line from P1 through P2; where this direction ray intersects P2's Ring is the ZERO point of that Ring. So the Direction of the Ring is dependent on the direction of the line to be displayed.

In the DCC only the differences between points on the consecutive Rings are coded. The position of point P3, i.e., the increment on the new Ring (centre P2) is described by the difference between the position of point P2 on the previous Ring and the position of the new point P3 on the current Ring, the positions numbered with respect to the previous Ring. As shown in the figure below, the position of point P3 is defined by the difference: P3 - P2 = -1. P3 and P2 being point numbers on the two Rings, numbered as given in the previous figure. The Direction (position of the point with number ZERO) is identified by D.

Change of direction with R=3

The basic Radius of the Ring, as used in Incremental mode, is dependent on the BGU. The BGU is the smallest nonzero value that can be expressed within the precision set by the VDC INTEGER or REAL PRECISION element. For VDC TYPE integer, basic Radius (=BGU) = 1. For VDC TYPE real the size of the BGU is set by a parameter of VDC REAL PRECISION, called smallest-real code: letting "src" stand for smallest-real code, $BGU=2^{src}$. The basic Radius is a multiple of the BGU, i.e. it is a positive integer. For VDC TYPE real the default value for the basic Radius follows from:

smallest-real code	basic Radius
$> g$	1
$= g$	
$< g$	$\dfrac{\frac{1}{2^g}}{BGU}$

where g is a nominal smallest-real code. A value of $g = -8$ is used in this Standard.

The basic Radius and the Angular resolution factor may be changed with the DOMAIN RING element. If the basic Radius is set to a value less than ONE, the value ONE is assumed for the basic Radius.

The encoding used in Incremental Mode makes use of the DCC property by using variable length codewords (Huffman Code). The encoding also allows changing of the Radius and the Angular resolution factor. The Radius can have a value of R, 2R, 4R or 8R, where R is the defined Radius. The Angular resolution factor p can be 0, 1, 2, or 3, the default value is 0.

The Huffman Code table used in the Incremental mode is a fixed length table. To allow the encoding of more points on a Ring two Escape codes are defined. With these Escape codes the points outside the Huffman Code table can be addressed. The end of the Incremental mode data is indicated by an End of Block value in the Huffman Code table.

The fixed Huffman Code table is given in table 2.

TABLE 2. Huffman Code table for Incremental mode.

Length	Code-word	Point number
2	00	0
2	10	1
2	01	-1
4	1100	2
4	1101	-2
6	111000	3
6	111001	-3
6	111010	4
6	111011	-4
8	11110000	5
8	11110001	-5
8	11110010	6
8	11110011	-6
8	11110100	7
8	11110101	-7
8	11110110	8
8	11110111	-8
10	1111100000	9
10	1111100001	-9
10	1111100010	10
10	1111100011	-10
10	1111100100	11
10	1111100101	-11
10	1111100110	12
10	1111100111	-12
10	1111101000	13
10	1111101001	-13
10	1111101010	14
10	1111101011	-14
10	1111101100	15
10	1111101101	-15
10	1111101110	16
10	1111101111	-16
10	1111110000	17
10	1111110001	-17
10	1111110010	18
10	1111110011	-18
10	1111110100	19
10	1111110101	-19
10	1111110110	C1
10	1111110111	-20
10	1111111000	C2
10	1111111001	C3
10	1111111010	C4
10	1111111011	C5
10	1111111100	C6
10	1111111101	IM-ESC 1
10	1111111110	IM-ESC 2
10	1111111111	End of Block

The <End of Block> code from the Huffman Code table identifies the end of the Incremental mode data. Remaining bits in the last Incremental mode data byte have no meaning; they are required to be zero.

The Incremental Mode escape codes <IM-ESC 1> and <IM-ESC 2> are used to extend the addressable number of points, e.g. points outside the range -20 to 19. The code <IM-ESC 1> adds +20 or -20 to the following code, depending on the sign of that following point. The code <IM-ESC 2> adds +40 or -40 to the following code, depending on the sign. The escape codes can follow each other in any desired order. The following examples demonstrate some possible combinations, [n] is a point number.

	<IM-ESC 1>	[1]	= point number 21
	<IM-ESC 1>	[-1]	= point number -21
	<IM-ESC 2>	[14]	= point number 54
	<IM-ESC 2>	[-12]	= point number -52
<IM-ESC 1>	<IM-ESC 2>	[6]	= point number (20+40+6) = 66
<IM-ESC 2>	<IM-ESC 1>	[-18]	= point number (-40-20-18) = -78

The codes C1 up to C6 are used to change the parameters that define the Ring to be used. The values of R are taken from the range R0, 2R0, 4R0 and 8R0, where R0 is the value of the Ring Radius before entering the Incremental Mode. The values of p are taken from the range 0, 1, 2 and 3. The function of these codes is as follows:

C1   Change the Ring parameters, R and p, to the next higher value, e.g. if the Radius is R0 it is set to 2R0, if the Angular resolution is p=0 it is set to p=1. R cannot become greater than 8R0 and p cannot become greater than 3. For example if the current Ring Radius is 8R0 and the current Angular resolution factor 3, the code <C1> has no effect.

C2   Change the Ring parameters, R and p, to the next lower value. The effect of the code <C2> is the inverse of code <C1>. R cannot become smaller than R0 and p cannot become smaller than 0. For example if the current Radius is R0 and the current Angular resolution factor 0, the code <C2> has no effect.

C3   Change the Ring Radius R to the next higher value. The code <C3> has no effect if the current Radius is 8R0.

C4   Change the Angular resolution factor p to the next higher value. The code <C4> has no effect if the current factor is 3.

C5   Change the Ring Radius R to the next lower value. The code <C5> has no effect if the current Radius is R0.

C6   Change the Angular resolution factor p to the next lower value. The code <C6> has no effect if the current factor is 0.

In addition, those codes (C1 to C6) set the position of the point with number ZERO on the positive x-axis, while the cartesian axes are drawn through the centre point of the Ring.

6.6.3 Incremental Mode Encoding

The structure of the incremental mode format is given below:

```
b8                 b1
+-+-+-+-+-+-+-+-+
|X|1|0|1|0|0|0|0|  first byte
+-+-+-+-+-+-+-+-+

+-+-+-+-+-+-+-+-+
|X|1|0|0|0|0|0|1|  second byte
+-+-+-+-+-+-+-+-+

+-+-+------------+
|X|1|b|b|b|b|b|b|  other bytes
+-+-+------------+
```

The first byte is coded according to the Basic format structure indicating an integer value "minus zero".

The second byte is structured according to the basic format structure. Bit b1 of the second byte is set to 1 to identify the use of Direct Chain Coding (DCC). The bits b5 to b2 are reserved for future use and shall be set to zero. The following bytes are coded according to the Bitstream format (see 6.2).

Because the Incremental mode uses variable length code words, these may not fit in the Incremental mode data bits (bits b6 to b1 of the other bytes). The code words are packed in consecutive bits of the Incremental mode bytes, starting from high numbered bits to lower numbered bits.

The end of Incremental mode data is identified by the <End of Block> code. Remaining bits in the last Incremental mode data byte have no meaning; they are required to be zero.

6.7 Colour Specifiers

Colour specifiers are coded either as colour indexes (if COLOUR SELECTION MODE is 'indexed') or as RGB parameters (if COLOUR SELECTION MODE is 'direct').

A colour index parameter is coded as an integer in the Basic format.

The RGB parameters used if COLOUR SELECTION MODE is 'direct' are coded as a series of bytes in the range of 4/0 to 7/15 (Bitstream format). The six least significant bits in each byte hold binary bits representing the red, green, and blue colour values, starting from high numbered bits to lower numbered bits from the first byte to the last byte. The number of bits needed to encode this data is set by COLOUR PRECISION. If this value is set to N bits, there are 3N colour value bits; i.e. there are only as many bytes in an RGB parameter as are necessary to hold those 3N bits. For example, if COLOUR PRECISION is set to 5 bits, RGB parameters have the following form:

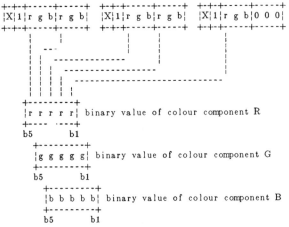

Each byte contains two bits for each component (R,G,B). The first byte contains the most significant bits. If bits b3 to b1 in the last byte are unused they are required to be zero.

6.8 Colour Lists

A colour list is a structured data type consisting of a coding type indicator followed by a sequence of colour list elements, each element specifying a colour value (indexed or direct).

Seven different formats are possible for colour lists, based on the current value of the COLOUR SELECTION MODE and the coding type indicator which precedes the colour values.

This coding type indicator is coded in Basic format and may have values of 0 through 3; these can be viewed as setting two one-bit flags. (Note: bits b5-b3 are reserved for future use, and shall be set to 0.) The coding type indicator:

```
     b8                 b1
     +-+-+-+---------+
     |X|1|0|0 0 0  r  f|
     +-+-+-+---------+
```

coding type	runlength(r)	format(f)	"name"
0	0	0	normal
1	0	1	bitstream
2	1	0	runlength
3	1	1	runlength bitstream

Runlength encoding is efficient if many adjacent colour list elements have the same colour value. For each run, the colour of the cells is specified, followed by a count telling how many cells are in the run. This count is always in Basic format; the format of the colour values is described below.

If adjacent colour cells are not the same colour, runlength encoding is less efficient than the normal or bitstream formats.

In the normal formats, colour values are coded as for individual colour parameters (see 6.7).

In the bitstream formats, colour values are packed together as tightly as possible, six bits at a time, in consecutive data bits starting from high-numbered bits to lower-numbered bits, into characters from columns 4 to 7 of the code chart. The parameters are permitted to straddle character boundaries.

Bitstream format parameters are not self-delimiting (see 6.2). Any unused bits in the last character of a Bitstream format parameter have no meaning; they are required to be zero.

6.8.1 Normal Format (coding type=0)

When COLOUR SELECTION MODE is 'indexed', the colour list is a sequence of numbers in Basic format, each an individual CI.

When COLOUR SELECTION MODE is 'direct', the colour list is a sequence of bitstreams, each bitstream representing one RGB value.

The normal format is the easiest to produce, and matches the formats used for individual colour values (e.g., LINE COLOUR).

6.8.2 Bitstream Format (coding type=1)

When COLOUR SELECTION MODE is 'indexed', the colour list is a single bitstream which is the concatenation of numbers representing CI in the Local Colour Precision. This form is similar to that used for "raster screen dumps" in many applications.

When COLOUR SELECTION MODE is 'direct', the colour list is a single bitstream which is the concatenation of the RGB bitstreams in the Local Colour Precision. This is the most compact form of direct colour list where colours vary greatly within a row.

6.8.3 Runlength Format (coding type=2)

When COLOUR SELECTION MODE is 'indexed', the colour list is a sequence of pairs, each pair consisting of [CI in Basic format at Local Colour Precision; number of repetitions in Basic format]. This is the simplest form of runlength indexed list.

When COLOUR SELECTION MODE is 'direct', the colour list is a sequence of pairs, each pair consisting of [RGB bitstream at Local Colour Precision; number of repetitions in Basic format].

6.8.4 Runlength Bitstream Format (coding type=3)

When COLOUR SELECTION MODE is 'indexed', the colour list is a sequence of pairs, each pair consisting of [CI in Bitstream format at Local Colour Precision; number of repetitions in Basic format]. The advantage over Runlength format is that the "extend bit" and "sign bit" of the Basic format are available as data bits for the index itself, which means that CI in the range 16..63 may be coded in one character instead of two.

When COLOUR SELECTION MODE is 'direct', this format is identical to the Runlength format.

6.8.5 Examples

The following example shows a colour index list in Runlength coding:

```
Coding:             Runlength
Datatype:           Colour Index List
Parameter Values:   2,3,3,3,3,4

4/2       first byte: r=1,f=0
4/2,4/1   Run Colour 1 (value 2), Run Count 1 (value 1)
4/3,4/4   Run Colour 2 (value 3), Run Count 2 (value 4)
4/4,4/1   Run Colour 3 (value 4), Run count 3 (value 1)
```

As an example for a colour indexed list, suppose that COLOUR INDEX PRECISION is 5 so that each colour index has 5 bits. Then, the colour indexes are packed into the Bitstream as follows:

```
+-+-----------+
|1|A A A A A B|
+-+-----------+
+-+-----------+
|1|B B B B C C|
+-+-----------+
+-+-----------+
|1|C C C D D D|
+-+-----------+
+-+-----------+
|1|D D 0 0 0 0|
+-+-----------+
```

Note that AAAAA is the binary numeral for the first colour index, BBBBB is the binary numeral for the second colour index, and so on.

The leftover bits in the last byte have no meaning; they are required to be zero.

The following example shows the Bitstream coding with the same colour index list as used in the Runlength coding:

Coding: Bitstream
Datatype: Colour Index List
Parameter Values: 2,3,3,3,3,4 (COLOUR INDEX PRECISION = 4)

4/1   first byte : r=0,f=1
4/8   Colour Index 1 and (half of) 2
7/3   Colour Index (half of) 2 and 3
4/12   Colour Index 4 and (half of) 5
7/4   Colour Index (half of) 5 and 6

In the Basic format the same colour index list would occupy more bytes, as shown in the following example:

Coding: Basic Fomat
Datatype: Colour Index List
Parameter Values: 2,3,3,3,3,4

4/0   first byte : r=0,f=0
4/2   Colour Index 1 (value 2)
4/3   Colour Index 2 (value 3)
4/3   Colour Index 3 (value 3)
4/3   Colour Index 4 (value 3)
4/3   Colour Index 5 (value 3)
4/4   Colour Index 6 (value 4)

6.9 String Parameters

6.9.1 Overall String Parameter Format

Strings are coded as sequences of bytes, starting with START OF STRING (SOS) and terminated by STRING TERMINATOR (ST).

SOS is represented by the C1 control 9/8. In a 7-bit environment SOS is coded ESC 5/8 (where ESC is the bit combination 1/11). ST is coded as ESC 5/12 in a 7-bit environment, and 9/12 in an 8-bit environment.

	7-bit environment	8-bit environment
SOS	1/11 5/8	09/08
ST	1/11 5/12	09/12

6.9.2 Bit Combinations Permitted within String Parameters of Text Elements

The following bit combinations are allowed as bytes in string parameters in a conforming metafile:

a) Bit combinations from columns 2 through 7 of a 7-bit or 8-bit code chart. If the CHARACTER CODING ANNOUNCER has selected an 8-bit coding technique, then bit combinations from columns 10 through 15 of an 8-bit code chart are also permitted within string parameters.

b) The following C0 control characters: NUL (0/0), BS (0/8), HT (0/9), LF (0/10), VT (0/11), FF (0/12), CR (0/13), SO (0/14), SI (0/15) and ESC (1/11). The NUL control is permitted, but has no effect on the meaning of the string. The bit combinations 0/8 to 0/13 are permitted, but have no standardized effect. The SO, SI, and ESC controls are permitted within conforming metafiles, dependent on the CHARACTER CODING ANNOUNCER. If the bit combinations 0/14 or 0/15 are present, they shall have the meanings (SO or LS1, and SI or LS0) specified for them in ISO 2022, unless a private coding technique has been selected by CHARACTER CODING ANNOUNCER. Other C0 controls are reserved for future standardization.

c) The ESCAPE control (bit combination 1/11 with the exception of the bit combinations 1/11 5/8 and 1/11 5/12 for SOS and ST), and escape sequences formed according to the rules specifed in ISO 2022. Those escape sequences for which meanings are specified in ISO 2022 shall have those specified meanings. ESCAPE sequences are only permitted if the CHARACTER CODING ANNOUNCER has selected a 2022-compatible technique. (If a private value of CHARACTER CODING ANNOUNCER is selected, ESCAPE may be used for escape sequences in a manner agreed upon by the interchanging parties. However this is not recommended as it will decrease the portability of the metafile.)

d) By agreement between interchanging parties, C1 controls, with the exception of the bit combinations SOS and ST for the string delimiter. (In an 8-bit environment, these are bit combinations from columns 8 and 9 of the code chart.) In a 7-bit environment, they are ESC Fe sequences, where ESC is the bit combination 1/11 and Fe is a bit combination from column 4 or 5 of the code chart. The bit combinations 08/14 and 08/15 (or ESC 4/14 and ESC 4/15), if present, shall have the meanings SS2 (SINGLE SHIFT TWO) and SS3 (SINGLE SHIFT THREE) as defined in ISO 2022 when the extended 7-bit or extended 8-bit mode has been selected using CHARACTER CODING ANNOUNCER.

The bit combinations 0/1 through 0/7, 1/0 through 1/10, and 1/12 through 1/15 are reserved for future standardization.

6.9.3 C0 Control within String Parameters

TABLE 3. C0 Control Set

Column 0			Column 1		
0/0	NUL	(NULL-ignored)	1/0	DLE	(reserved)
0/1	SOH	(reserved)	1/1	DC1	(reserved)
0/2	STX	(reserved)	1/2	DC2	(reserved)
0/3	ETX	(reserved)	1/3	DC3	(reserved)
0/4	EOT	(reserved)	1/4	DC4	(reserved)
0/5	ENQ	(reserved)	1/5	NAK	(reserved)
0/6	ACK	(reserved)	1/6	SYN	(reserved)
0/7	BEL	(reserved)	1/7	ETB	(reserved)
0/8	BS	(no standardized effect)	1/8	CAN	(reserved)
0/9	HT	(no standardized effect)	1/9	EM	(reserved)
0/10	LF	(no standardized effect)	1/10	SUB	(reserved)
0/11	VT	(no standardized effect)	1/11	ESC	(ESCAPE)
0/12	FF	(no standardized effect)	1/12	IS4	(reserved)
0/13	CR	(no standardized effect)	1/13	IS3	(reserved)
0/14	SO	(SHIFT OUT)	1/14	IS2	(reserved)
0/15	SI	(SHIFT IN	1/15	IS1	(reserved)

6.9.4 Using G-Sets in String Parameters

The G-sets (G0, G1, G2, and G3) are coded character sets of 94 or 96 bit combinations. They may be invoked into columns 2 through 7 of a 7-bit code chart, or columns 02 through 07 and 10 through 15 of an 8-bit code chart. This encoding of the CGM uses G-sets within string parameters, as follows:

6.9.4.1 String Parameters of TEXT, APPEND TEXT, and RESTRICTED TEXT. For the string parameters of TEXT, APPEND TEXT, and RESTRICTED TEXT, the CHARACTER SET INDEX element designates a particular character set (from the list established by CHARACTER SET LIST) as the G0 set. Likewise, ALTERNATE CHARACTER SET INDEX designates a particular character set as both the G1 and G2 sets. These attributes may also apply to string parameters within data records of GDP elements.

At the start of each picture, BEGIN PICTURE invokes the G0 set into columns 2 through 7 of a 7-bit or 8-bit code chart. In an 8-bit environment, BEGIN PICTURE invokes the G1 set into columns 10 through 15 of the 8-bit code chart.

The ISO 2022 C0 or C1 controls or escape sequences to designate character sets as G0, G1, G2, or G3, or to invoke G-sets into the 7-bit or 8-bit code chart ("in-use table") may be included within the string parameters of the graphical primitive elements identified in 6.9.4.1 (with the exception of the string delimiter, listed in 6.9.1), if the CHARACTER CODING ANNOUNCER has selected an extended 2022 technique, whereupon they assume the meanings specified in ISO 2022.

6.9.4.2 String Parameters of Other Metafile Elements. The CHARACTER SET INDEX and ALTERNATE CHARACTER SET INDEX elements apply only to the string parameters of the graphical primitive elements identified in 6.9.4.1. String parameters of other metafile elements are not part of the pictures stored in the metafile. This Standard does not specify the effect, if any, of using characters from code chart columns 10 through 15 in the string parameters of metafile elements other than the graphical primitive elements identified in 6.9.4.1, nor will any future revision of this Standard. This Standard does not specify the effect of ISO 2022 designating and invoking controls when they occur within the string parameters of metafile elements other than the graphical primitive elements identified in 6.9.4.1.

6.10 Enumerated Parameters

Enumerated parameters represent choices within a fixed set of standardized options. All of the enumerated types may have private values. They are coded as integers (Basic format). Private (non-standard) values of enumerated parameters shall use negative integers. Some but not all of these have been shown in the encodings in clause 8 below.

6.11 Index Parameters

Indexes are coded as integers (Basic format), at INDEX PRECISION. Private (non-standard) values of index parameters are all coded using negative integers.

6.12 Data Record Parameters

Data record parameters are coded identically to string parameters in this encoding.

7 Character Substitution

To accommodate systems in which it is inconvenient or impossible to include C0 control characters, the SPACE character (2/0), or the DELETE character (7/15) in the metafile, this CGM encoding includes a "character substitution" option. Characters in the range of 0/0 to 1/15, the characters 2/0, 7/14, and 7/15 may be replaced by 2-byte sequences provided each such 2-byte sequence is declared in the first parameter of the BEGIN METAFILE element.

Although the first parameter, 'substitution strings', of the BEGIN METAFILE element is of data type "string" the START OF STRING (SOS) introducer is omitted in this special case. This is because the SOS contains characters for which character substitution may be required. The STRING TERMINA-TOR (ST), however, is included in this parameter (character substitution, if selected, will already be in effect by the end of the string parameter).

Table 4 shows the characters that may be replaced by such 2-byte sequences, and the 2-byte sequences with which they are replaced. Note that all the 2-byte sequences begin with 7/14. Therefore, if the character substitution option is used at all, that character (7/14) shall be declared as one of the characters that is being replaced with a 2-byte sequence. This is done by including its replacement sequence (7/14 3/14) in the first parameter of BEGIN METAFILE.

Once a character (or rather, its 2-byte replacement) has been declared in the BEGIN METAFILE element, "character substitution" is in effect immediately for that character and will remain in effect until the end of the metafile (that is, until the END METAFILE element). The metafile interpreter would ignore any characters for which character substitution is in effect.

Example:

Supposing that it is desirable to avoid using the characters SPACE, TILDE, and DELETE in the metafile, and for the metafile interpreter to ignore those characters if they are inadvertently inserted (for example, by a host operating system or some process other than the metafile generator).

In that case, the metafile generator declares "character substitution" for the above two characters and for the TILDE character, 7/14. It does this in the first parameter of BEGIN METAFILE as follows:

```
BEGIN METAFILE
   = 3/0  2/0                    {BEGIN METAFILE opcode}

   7/14 6/0  7/14 3/14 7/14 3/15    { string: ˜‘˜>˜?}
   1/11 5/12                     { STRING TERMINATOR (ST)}

   1/11 5/8                      { START OF STRING (SOS)}
   metafile name
   1/11 5/12                     { ST}
```

Throughout the metafile, wherever the metafile generator would otherwise put a SPACE, TILDE (7/14), or DELETE character, it substitutes the 2-byte sequences 7/14 6/0, 7/14 3/14, or 7/14 3/15, respectively.

In the preceding example, wherever the metafile interpreter encounters the character 7/14, it interprets it as the first character of a 2-byte sequence representing one of these characters. The metafile interpreter would then ignore the characters 2/0 (SPACE), and 7/15 (DELETE).

TABLE 4. Character Substitution

The character:		May be replaced with:		ISO 646
		7-bit	8-bit	char.
0/0	(NUL)	7/14 4/0	07/14 04/0	(¯@)
0/1	(SOH)	7/14 4/1	07/14 04/1	(¯A)
0/2	(STX)	7/14 4/2	07/14 04/2	(¯B)
0/3	(ETX)	7/14 4/3	07/14 04/3	(¯C)
0/4	(EOT)	7/14 4/4	07/14 04/4	(¯D)
0/5	(ENQ)	7/14 4/5	07/14 04/5	(¯E)
0/6	(ACK)	7/14 4/6	07/14 04/6	(¯F)
0/7	(BEL)	7/14 4/7	07/14 04/7	(¯G)
0/8	(BS)	7/14 4/8	07/14 04/8	(¯H)
0/9	(HT)	7/14 4/9	07/14 04/9	(¯I)
0/10	(LF)	7/14 4/10	07/14 04/10	(¯J)
0/11	(VT)	7/14 4/11	07/14 04/11	(¯K)
0/12	(FF)	7/14 4/12	07/14 04/12	(¯L)
0/13	(CR)	7/14 4/13	07/14 04/13	(¯M)
0/14	(SO)	7/14 4/14	07/14 04/14	(¯N)
0/15	(SI)	7/14 4/15	07/14 04/15	(¯O)
1/0	(DLE)	7/14 5/0	07/14 05/0	(¯P)
1/1	(DC1)	7/14 5/1	07/14 05/1	(¯Q)
1/2	(DC2)	7/14 5/2	07/14 05/2	(¯R)
1/3	(DC3)	7/14 5/3	07/14 05/3	(¯S)
1/4	(DC4)	7/14 5/4	07/14 05/4	(¯T)
1/5	(NAK)	7/14 5/5	07/14 05/5	(¯U)
1/6	(SYN)	7/14 5/6	07/14 05/6	(¯V)
1/7	(ETB)	7/14 5/7	07/14 05/7	(¯W)
1/8	(CAN)	7/14 5/8	07/14 05/8	(¯X)
1/9	(EM)	7/14 5/9	07/14 05/9	(¯Y)
1/10	(SUB)	7/14 5/10	07/14 05/10	(¯Z)
1/11	(ESC)	7/14 5/11	07/14 05/11	(¯[)
1/12	(FS or IS4)	7/14 5/12	07/14 05/12	(¯\)
1/13	(GS or IS3)	7/14 5/13	07/14 05/13	(¯])
1/14	(RS or IS2)	7/14 5/14	07/14 05/14	(¯^)
1/15	(US or IS1)	7/14 5/15	07/14 05/15	(¯_)
2/0	(SPACE)	7/14 6/0	07/14 06/0	(¯`)
7/14	(TILDE,¯)	7/14 3/14	07/14 03/14	(¯>)
7/15	(DELETE)	7/14 3/15	07/14 03/15	(¯?)

In a 7-bit or an 8-bit CGM coding environment the "GL" part of the code table (columns 02 through 07), table 4 may be summarized as follows: each character from decimal 0 (0/0) to decimal 32 (2/0) may be replaced by a 2-byte sequence in which the first byte is decimal 126 (7/14) and the decimal equivalent of the second byte is obtained by adding 64, modulo 128, to the decimal equivalent of the original character.

8 Representation of Each Element

For convenience, the 7-bit opcode is given for each element described in the remainder of this clause. To determine the 8-bit opcode, a zero is added in front of the column specification (for example, 02/1 instead of 2/1). A list of opcodes is given in table 1.

Notation Used:

<symbol>*	=	0 or more occurrences
<symbol>+	=	1 or more occurences
<symbol>o	=	optional, 0 or more occurences
{comment}	=	explanation of a production
<x:y>	=	construct x with meaning y

8.1 Delimiter Elements

8.1.1 BEGIN METAFILE

<BEGIN-METAFILE-opcode: 3/0 2/0>
<string: substitution-codes>
<string: metafile-identifier>

The BEGIN METAFILE and END METAFILE elements respectively mark the beginning and the end of a metafile. (More than one metafile can be recorded in a single computer file provided each metafile is delimited by the elements BEGIN METAFILE and END METAFILE.)

The first parameter, <string: substitution-codes>, lists those 2-character substitution codes that will be used in the metafile to represent C0 control characters (bit combinations in the range of 0/0 to 1/15) or the SPACE, TILDE, and DELETE characters (bit combinations 2/0, 7/14, and 7/15, respectively. Character substitution will be in effect only for those bit combinations whose substitution codes are listed in the first <string> parameter. If the first <string> parameter is empty (represented by a single ST), character substitution is not to be used in this metafile.

NOTE - Substitution codes are not opened by an SOS introducer, see clause 7.

8.1.2 END METAFILE

<END-METAFILE-opcode: 3/0 2/1>

8.1.3 BEGIN PICTURE

<BEGIN-PICTURE-opcode: 3/0 2/2>
<string: picture-identifier>

8.1.4 BEGIN PICTURE BODY

<BEGIN-PICTURE-BODY-opcode: 3/0 2/3>

8.1.5 END PICTURE

<END-PICTURE-opcode: 3/0 2/4>

8.2 Metafile Descriptor Elements

8.2.1 METAFILE VERSION

<METAFILE-VERSION-opcode: 3/1 2/0>
<integer: version>

8.2.2 METAFILE DESCRIPTION

<METAFILE-DESCRIPTION-opcode: 3/1 2/1>
<string: description-of-the-metafile>

8.2.3 VDC TYPE

<VDC-TYPE-opcode: 3/1 2/2>
<enumerated: VDC-type>

<enumerated:VDC-type>   =   <integer: 0> {integer}
                                   ¦   <integer: 1> {real}

8.2.4 INTEGER PRECISION

<INTEGER-PRECISION-opcode: 3/1 2/3>
<integer: largest-integer-code + 1 >

The largest-integer-code tells how many bits occur in the largest possible magnitude for an integer. For example, if integers in the metafile can range from -32767 to +32767, the largest-integer-code is 15. One additional bit is required for the sign, and so is added to obtain the proper precision. Thus in this example the parameter would be 16.

8.2.5 REAL PRECISION

<REAL-PRECISION-opcode: 3/1 2/4>
<integer: largest-real-code+1>
<integer: smallest-real-code>
<integer: default-exponent-for-reals>
<enumerated: exponents-allowed>

<enumerated: exponents-allowed>   =   <integer: 0> {allowed}
                                   ¦   <integer: 1> {forbidden}

The largest-real-code and smallest-real-code together specify the maximum number of bits of precision that can occur in the coded representations of real numbers. The largest-real-code specifies the largest possible magnitude of positive or negative real numbers, while the smallest-real-code specifies the granularity of the real number space.

The largest-real-code tells how many bits can occur to the left of the (binary) radix point in the largest possible real number. For example, if the largest possible real number is binary +11111111111111111111.111, the largest-real-code would be 20.

The smallest-real-code shows the granularity of the real number space by indicating how small a non-zero real number can be. It does this by specifying the smallest exponent that is permitted in the coded representation of a real number. For example, if all real numbers are integer multiples of 1/64 (2 raised to the power -6), the smallest-real-code would be -6. This indicates that there are six bits of precision to the right of the binary radix point and that the smallest allowable exponent is -6.

It is possible for the smallest-real-code to be positive. For example, if the only real numbers to be coded are multiples of 8, smallest-real-code would be +3, since 8, or binary 1000, is 2 raised to the power +3.

The largest-real-code shall be greater than the smallest-real-code. The difference, largest-real-code minus smallest-real-code, gives the number of bits of precision which are required in order to store real numbers with full precision.

The default-exponent-for-reals is the exponent that is to be assumed if an exponent is omitted from the coded representation of a real number. This parameter shall be greater than or equal to the smallest-real-code.

The exponents-allowed parameter specifies whether or not an "exponent follows" bit is included in the mantissa of the coded representation of each number. If this parameter is 0, exponents are allowed in real parameters, and the exponent-follows bit is included in their mantissas. If this parameter is 1, exponents are not allowed in real parameters and the mantissas of real parameters do not include the exponent-follows bit. In that case, the omitted exponents of all real parameters are deemed to have the value determined by the default-exponent-for-reals parameter.

8.2.6 INDEX PRECISION

<INDEX-PRECISION-opcode: 3/1 2/5>
<integer: largest-index-code + 1>

The largest-index-code tells how many bits occur in the largest possible magnitude for an index. For example, if indexes in the metafile can range from -511 to +511, the largest-integer-code is 9. One additional bit is required for the sign, and so is added to obtain the proper precision. Thus in this example the parameter would be 16.

8.2.7 COLOUR PRECISION

<COLOUR-PRECISION-opcode: 3/1 2/6>
<integer: largest-component code>

The largest-component-code tells how many bits occur in the largest possible magnitude for a direct colour component, i.e., for each of the red, green, and blue components. For example, if the largest-component-code is 10, then direct colour components in the range 0-1023 can be handled.

8.2.8 COLOUR INDEX PRECISION

<COLOUR-INDEX-PRECISION-opcode: 3/1 2/7>
<integer: largest-colour-index-code+1>

The parameter specifies the number of bits used in the binary numerals that represent colour indexes. For example, if the parameter is 4, then the largest-colour-index-code is 3, and there are 8 possible colour indexes: binary 0000 (decimal 0) to binary 0111 (decimal 7).

8.2.9 MAXIMUM COLOUR INDEX

<MAXIMUM-COLOUR-INDEX-opcode: 3/1 2/8>
<integer: maximum-colour-index>

8.2.10 COLOUR VALUE EXTENT

<COLOUR-VALUE-EXTENT-opcode: 3/1 2/9>
<RGB: minimum-colour-value>
<RGB: maximum-colour-value>

8.2.11 METAFILE ELEMENT LIST

<METAFILE-ELEMENT-LIST-opcode: 3/1 2/10>
<string: element-list>

<string: element-list>	=	SOS <element names>* ST
<element names>	=	<element opcode>
	¦	<enumerated: element set>
<enumerated: element set>	=	<integer: 0> {DRAWING SET}
	¦	<integer: 1> {DRAWING PLUS CONTROL SET}
	¦	<integer: negative> {privately defined set}

8.2.12 METAFILE DEFAULTS REPLACEMENT

<BEGIN-METAFILE-DEFAULTS-opcode: 3/1 2/11>
<element-that-sets-a-default>+
<END-METAFILE-DEFAULTS-opcode: 3/1 2/12>

The <element-that-sets-a-default> list is given in part 1, clause 6. The coding format for each of the elements that appear in METAFILE DEFAULTS REPLACEMENT is exactly as described in this clause, Representation of Each Element. The parameter values express the default values to be used. For example, in a 7-bit encoding, the default values for CLIP INDICATOR and HATCH INDEX would be as follows:

3/1 2/11	{BEGIN-METAFILE-DEFAULTS-opcode}
3/3 2/5	{CLIP-INDICATOR-opcode}
integer: 1	{clip-indicator-on}
3/6 3/3	{HATCH-INDEX-opcode}
integer: 2	{default index equal to 2}
3/1 2/12	{END-METAFILE-DEFAULTS-opcode}

8.2.13 FONT LIST

<FONT-LIST-opcode: 3/1 2/13>
<font-declaration>+

<font-declaration>	=	<string: name-of-font>

The FONT LIST element declares the character fonts that may be named in subsequent TEXT FONT INDEX elements and establishes the font index value that is associated with each such character font.

The first <font-declaration> in the list names the character font whose font index value is to be 1. Likewise, the second, third, fourth, etc., <font-declaration>s name the fonts whose index values are to be 2, 3, 4, etc. The <font-declaration>s are separated from each other by the string delimiter characters.

8.2.14 CHARACTER SET LIST

<CHARACTER-SET-LIST-opcode: 3/1 2/14>
<character-set-declaration>+

<character-set-declaration>   =   <enumerated: character-set-type>
                                          <string: end-of-escape-sequence-to-designate-set>

<enumerated: character-set-type>   =   <integer: 0> {94-character G-set}
                                     ¦   <integer: 1> {96-character G-set}
                                     ¦   <integer: 2> {multibyte 94-character G-set}
                                     ¦   <integer: 3> {multibyte 96-character G-set}
                                     ¦   <integer: 4> {"complete code" character set}

The CHARACTER SET LIST element declares the character sets that can be named in subsequent
CHARACTER SET INDEX and ALTERNATE CHARACTER SET INDEX elements and establishes
the character set index value that is associated with each of these character sets.

The first <character-set-declaration> in the list names the character set whose character set index
value is to be 1. Likewise, the second, third, fourth, etc., <character-set-declaration>s name the
character sets whose index values are to be 2, 3, 4, etc.

Each <character-set-declaration> has two parts: an enumerated <integer> and a short <string>.
The <integer> specifies which type of character set is being declared (that is, which type of ISO
2022 designating escape sequence is associated with that character set). The <string> consists of
the character that forms the "tail end" of such designating escape sequences for that character set.

There are five types of character sets: 94-character G-sets, 96-character G-sets, 94-character multibyte G-sets, 96-character multibyte G-sets, and character sets intended to be designated as "complete codes". Examples given below are for an 8-bit environment.

8.2.14.1 94-CHARACTER G-SETS. A CHARACTER SET LIST element could specify that the U.K.
character set is to be referred to by character set index 1, and the French character set by character set
index 2, as follows:

>   <CHARACTER-SET-LIST-opcode>
>   <integer: 0> SOS 4/1 ST
>   <integer: 0> SOS 6/6 ST

>   = 03/1 02/14 04/0 09/8 04/1 09/12 04/0 09/8 06/6 09/12

8.2.14.2 96-CHARACTER G-SETS. The following CHARACTER SET LIST element establishes the
U.K. 94-character G-set, the French 94-character G-set, and a private 96-character G-set as the character sets named by character set indexes 1, 2, and 3, respectively:

>   <CHARACTER-SET-LIST-opcode>
>   <integer: 0> SOS 4/1 ST
>   <integer: 0> SOS 6/6 ST
>   <integer: 1> SOS 3/0 ST

>   = 03/1 02/14 04/0 09/8 04/1 09/12
>   04/0 09/8 06/6 09/12 04/1 09/8 03/0 09/12

8.2.14.3 94-CHARACTER MULTIBYTE G-SETS. A Japanese 2-byte character set of 6802 graphic
characters has been registered in the International Register Of Character Sets To Be Used With Escape
Sequences, and its designating escape sequences have the form shown above, with the final character
<F> being 4/0. Thus, the following CHARACTER SET LIST element could be used to specify that
this 2-byte Japanese character set is to be referred to by character set index 2:

>   <CHARACTER-SET-LIST-opcode>

<integer: 2> SOS 4/0 ST

= 03/1 02/14 04/2 09/8 04/0 09/12

8.2.14.4 96-CHARACTER MULTIBYTE G-SETS OF GRAPHIC CHARACTERS. It is not possible to designate a 96-character multibyte G-set as a G0 set.

The <character-set-declaration> for a 96-character multibyte G-set consists of <integer: 3> followed by a <string> consisting only of the final character <F> in the character set's ISO 2022 designating escape sequence.

So far, no 96-character multibyte G-sets have been registered in the International Register of Character Sets To Be Used with Escape Sequences.

8.2.14.5 CHARACTER SETS INTENDED TO BE DESIGNATED AS COMPLETE CODES. The following CHARACTER SET LIST element would declare the French character set to have character set index 1 and that 8-bit private code to have character set index 2:

<CHARACTER-SET-LIST-opcode>
<integer: 0> SOS 6/6 ST
<integer: 4> SOS 2/0 3/0 ST

= 03/1 02/14 04/0 09/8 06/6 09/12 04/4 09/8 02/0 03/0 09/12

8.2.15 CHARACTER CODING ANNOUNCER

<CHARACTER-CODING-ANNOUNCER-opcode: 3/1 2/15>
<enumerated: coding-technique>

<enumerated: coding-technique>	=	<integer: 0>	{basic 7-bit}
	¦	<integer: 1>	{basic 8-bit}
	¦	<integer: 2>	{extended 7-bit}
	¦	<integer: 3>	{extended 8-bit}
	¦	<integer: negative>	{private}

8.3 Picture Descriptor Elements

8.3.1 SCALING MODE

<SCALING-MODE-opcode: 3/2 2/0>
<enumerated: scaling-mode>
<real: metric-scale-factor>

<enumerated: scaling-mode>   =   <integer: 0> {abstract}
                                  ¦   <integer: 1> {metric}

8.3.2 COLOUR SELECTION MODE

<COLOUR-SELECTION-MODE-opcode: 3/2 2/1>
<enumerated: colour-selection-mode>

<enumerated: colour-selection-mode> =   <integer: 0> {indexed}
                                  ¦   <integer: 1> {direct}

8.3.3 LINE WIDTH SPECIFICATION MODE

<LINE-WIDTH-SPECIFICATION-MODE-opcode: 3/2 2/2>
<enumerated: specification-mode>

<enumerated: speci. ation-mode>   =   <integer: 0> {absolute}
                                  ¦   <integer; 1> {scaled}

8.3.4 MARKER SIZE SPECIFICATION MODE

<MARKER-SIZE-SPECIFICATION-MODE-opcode: 3/2 2/3>
<enumerated: specification-mode>

<enumerated: specification-mode> =   <integer: 0> {absolute}
                                  ¦   <integer: 1> {scaled}

8.3.5 EDGE WIDTH SPECIFICATION MODE

<EDGE-WIDTH-SPECIFICATION-MODE-opcode: 3/2 2/4>
<enumerated: specification-mode>

<enumerated: specification-mode> =   <integer: 0> {absolute}
                                  ¦   <integer: 1> {scaled}

8.3.6 VDC EXTENT

<VDC-EXTENT-opcode: 3/2 2/5>
<point: first-corner>
<point: second-corner>

8.3.7 BACKGROUND COLOUR

<BACKGROUND-COLOUR-opcode: 3/2 2/6>
<RGB: background colour>

8.4 Control Elements

8.4.1 VDC INTEGER PRECISION

<VDC-INTEGER-PRECISION-opcode: 3/3 2/0>
<integer: largest-integer-code+1>

The largest-integer-code tells how many bits occur in the largest possible magnitude for an integer point. For example, if integer points in the metafile can range from -32767 to +32767, the largest-integer-code is 15. One additional bit is required for the sign, thus an additional bit is added to obtain the appropiate field width. Thus in this example the parameter would be 16.

8.4.2 VDC REAL PRECISION

<VDC-REAL-PRECISION-opcode: 3/3 2/1>
<integer: largest-real-code+1>
<integer: smallest-real-code>
<integer: default-exponent-for-vdc-reals>
<enumerated: exponents-allowed>

<enumerated: exponents-allowed>   =   <integer: 0> {allowed}
                                     ¦   <integer: 1> {forbidden}

The largest-real-code and smallest-real-code (or granularity code) together specify the maximum number of bits of precision that can occur in the coded representations of real VDC space coordinates and VDC scalars. The largest-real-code specifies the largest possible magnitude of positive or negative real VDC coordinates and VDC scalars, while the smallest-real-code specifies the granularity of VDC space.

The largest-real-code tells how many bits can occur to the left of the (binary) radix point in the largest possible VDC scalar. For example, if the largest possible real number is binary +1111111111111111111.111, the largest-real-code would be 20.

The smallest-real-code shows the granularity of VDC space by indicating how small a non-zero VDC scalar can be. It does this by specifying the smallest exponent that is permitted in the coded representation of a VDC scalar, thus determining the Basic Grid Unit: letting "src" stand for smallest-real-code, $BGU=2^{src}$. For example, if all VDC scalars are integer multiples of 1/64 (2 raised to the power -6), the smallest-real-code would be -6. This indicates that there are six bits of precision to the right of the binary radix point and that the smallest allowable exponent is -6.

It is possible for the smallest-real-code to be positive. For example, if all VDC scalars to be coded are multiples of 8, smallest-real-code would be +3, since 8, or binary 1000, is 2 raised to the power +3.

The largest-real-code shall be greater than the smallest-real-code. The difference, largest-real-code minus smallest-real-code, gives the number of bits of precision which are required in order to store VDC coordinates with full precision. It is especially important to use all these bits of precision when coding the point list parameters of the POLYLINE, DISJOINT POLYLINE, POLYMARKER, POLYGON, POLYGON SET, and GDP.

The default-exponent-for-vdc-reals is the exponent that is to be assumed if an exponent is omitted from the coded representation of a real VDC coordinate or VDC scalar. This parameter shall be greater than or equal to the smallest-real-code.

The exponents-allowed parameter specifies whether or not an "exponent follows" bit is included in the mantissa of the coded representation of a real VDC coordinate or VDC scalar. If this parameter is 0, exponents are allowed in real VDC coordinates and VDC scalars, and the exponent-follows bit is included in their mantissas (see 6.5.) If this parameter is 1, exponents are not allowed in real VDC coordinates and VDC scalars; in that case the mantissas of real VDC coordinates and VDC scalar do not include the exponent-follows bit. In that case, the omitted exponents of all real parameters are deemed to have the value determined by the default-exponent-for-reals parameter.

8.4.3 AUXILIARY COLOUR

<AUXILIARY-COLOUR-opcode: 3/3 2/2>
<colour-specifier: auxiliary-colour>

<colour-specifier>   =   <integer: colour-index>   {if COLOUR SELECTION MODE is indexed}
                      ¦   <RGB> {if COLOUR SELECTION MODE is direct}

8.4.4 TRANSPARENCY

<TRANSPARENCY-opcode: 3/3 2/3>
<enumerated: transparency indicator>

<enumerated: transparency indicator>   =<integer: 0> {off}
                                        ¦   <integer: 1> {on}

8.4.5 CLIP RECTANGLE

<CLIP-RECTANGLE-opcode: 3/3 2/4>
<point: first-corner>
<point: second-corner>

8.4.6 CLIP INDICATOR

<CLIP-INDICATOR-opcode: 3/3 2/5>
<enumerated: clip-indicator>

<enumerated: clip-indicator>   =   <integer: 0> {off}
                                ¦   <integer: 1> {on}

8.5 Graphical Primitive Elements

8.5.1 POLYLINE

<POLYLINE-opcode: 2/0>
<point-list>

<point-list>   =   <point><point>+

8.5.2 DISJOINT POLYLINE

<DISJOINT-POLYLINE-opcode: 2/1>
<point-list>

8.5.3 POLYMARKER

<POLYMARKER-opcode: 2/2>
<point-list>

8.5.4 TEXT

<TEXT-opcode: 2/3>
<point: starting-point>
<enumerated: final-or-not-final>
<string: text-to-be-displayed>

<enumerated: final-or-not-final>   =   <integer: 0> {not final}
                                       ¦   <integer: 1> {final}

The string parameter may contain characters from G sets together with the ISO 2022 designating and/or invoking controls to select those G-sets as described below. Other C0 and C1 controls (except as noted in 6.9.2) may be included by agreement between exchanging parties but have no standardized meaning.

Within the string parameter, characters from columns 2 through 7 are displayed using the GL character set selected by the most recent CHARACTER SET INDEX element, the default character set, or by the most recent locking shift. If the character set selected by CHARACTER SET INDEX (or by default) is a 94-character G-set, the bit combinations 2/0 and 7/15 represent SPACE and DELETE, respectively.

Characters from columns 10 through 15 are displayed using the GR character set selected by the most recent ALTERNATE CHARACTER SET INDEX or locking shift right.

The ISO 2022 controls may occur within the string parameter provided their use is declared with the CHARACTER CODING ANNOUNCER element.

8.5.5 RESTRICTED TEXT

<RESTRICTED-TEXT-opcode: 2/4>
<VDC: delta-width>
<VDC: delta-height>
<point: starting-point>

<enumerated: final-or-not-final>
<string: text-to-be-displayed>

<enumerated: final-or-not-final>   =   <integer: 0> {not final}
                   ¦   <integer: 1> {final}

See 8.5.4, TEXT, for a description of the string parameter for RESTRICTED TEXT.

8.5.6 APPEND TEXT

<APPEND-TEXT-opcode: 2/5>
<enumerated: final-or-not-final>
<string: text-to-be-displayed>

<enumerated: final-or-not-final>   =   <integer: 0> {not final}
                   ¦   <integer: 1> {final}

See 8.5.4, TEXT, for a description of the parameters for APPEND TEXT.

8.5.7 POLYGON

<POLYGON-opcode: 2/6>
<point-list>

8.5.8 POLYGON SET

<POLYGON-opcode: <2/7>
<flagged point-list>

<flagged point-list>   =   <flagged-point>+

where the first coordinate is coded as an absolute and subsequent coordinates as incremental, as in the point list parameter type.

<flagged-point>   =   <coordinate> <enumerated: edge-out>

<enumerated: edge out>   =   <integer: 0>   {invisible}
                     ¦   <integer: 1>   {visible}
                     ¦   <integer: 2>   {close invisible}
                     ¦   <integer: 3>   {close visible}

8.5.9 CELL ARRAY

<CELL-ARRAY-opcode: 2/8>
<P: point>
<Q: point>
<R: point>
<integer: nx>
<integer: ny>
<local-colour-precision>
<colour-list>

<local-colour-precision>   =   <default-colour-precision-indicator>

		¦	<number-of-bits>
<default-colour-precision-indicator>	=		<integer: 0>
<number-of-bits>	=		<integer: positive>

The value 0 for the local-colour-precision is the 'default colour precision indicator'. It indicates that the colour specifications of the CELL ARRAY element are in the precision of the COLOUR INDEX PRECISION or the COLOUR PRECISION (depending upon the value of the COLOUR SELECTION MODE). If positive, then it is the 'number of bits' of the colour specifications of the CELL ARRAY element (as in COLOUR PRECISION and COLOUR INDEX PRECISION).

<colour-list>	=	<indexed colour list>
	¦	<direct colour list>
<indexed colour list>	=	<indexed normal colour list>
	¦	<indexed bitstream colour list>
	¦	<indexed runlength colour list>
	¦	<indexed runlength bitstream colour list>
<direct colour list>	=	<direct normal colour list>
	¦	<direct bitstream colour list>
	¦	<direct runlength colour list>
	¦	<direct runlength bitstream colour list>
<indexed normal colour list>	=	<integer: 0>
		<basic format>+
<indexed bitstream colour list>	=	<integer: 1>
		<bitstream: concatenation of colour indexes>
<indexed runlength colour list>	=	<integer:2>
		<indexed run>+
<indexed run>	=	<integer: colour index>
		<integer: number of repetitions>
<indexed runlength bitstream colour list>	=	<integer:3>
		<indexed bitstream run>+
<indexed bitstream run>	=	<bitstream: colour index>
		<integer: number of repetitions>
<direct normal colour list>	=	<integer: 0>
		<RGB: colour value>+
<direct bitstream colour list>	=	<integer: 1>
		<bitstream: concatenation of colour values>
<direct runlength colour list>	=	<integer: 2>
		<direct run>+
<direct run>	=	<RGB: colour value>
		<integer: number of repetitions>
<direct runlength bitstream colour list>	=	<integer: 3>

<direct run>+

8.5.10 GENERALIZED DRAWING PRIMITIVE

<GDP-opcode: 2/9>
<integer: GDP-identifier>
<point-list>
<data-record: data-record>

8.5.11 RECTANGLE

<RECTANGLE-opcode: 2/10>
<point: first-corner>
<point: second-corner>

8.5.12 CIRCLE

<CIRCLE-opcode: 3/4 2/0>
<point: centre-of-circle>
<VDC: radius-of-circle>

8.5.13 CIRCULAR ARC 3 POINT

<CIRCULAR-ARC-3-POINT-opcode: 3/4 2/1>
<point: starting-point>
<point: intermediate-point>
<point: ending-point>

8.5.14 CIRCULAR ARC 3 POINT CLOSE

<CIRCULAR-ARC-3-POINT-CLOSE-opcode: 3/4 2/2>
<point: starting-point>
<point: intermediate-point>
<point: ending-point>
<enumerated: close-type>

<enumerated: close-type>   =   <integer: 0>   {pie}
                                  ¦   <integer: 1>   {chord}

8.5.15 CIRCULAR ARC CENTRE

<CIRCULAR-ARC-CENTRE-opcode: 3/4 2/3>
<point: centrepoint>
<VDC: DX_start>
<VDC: DY_start>
<VDC: DX_end>
<VDC: DY_end>
<VDC: radius>

8.5.16 CIRCULAR ARC CENTRE CLOSE

<CIRCULAR-ARC-CENTRE-CLOSE-opcode: 3/4 2/4>
<point: centrepoint>
<VDC: DX_start>
<VDC: DY_start>
<VDC: DX_end>
<VDC: DY_end>
<VDC: radius>
<enumerated: close-type>

<enumerated: close-type>   =   <integer: 0>   {pie}
                                          | <integer: 1>   {chord}

8.5.17 ELLIPSE

<ELLIPSE-opcode: 3/4 2/5>
<point: centrepoint>
<point: endpoint of first conjugate diameter>
<point: endpoint of second conjugate diameter>

8.5.18 ELLIPTICAL ARC

<ELLIPTICAL-ARC-opcode: 3/4 2/6>
<point: centrepoint>
<point: endpoint of first conjugate diameter>
<point: endpoint of second conjugate diameter>
<VDC: DX_start>
<VDC: DY_start>
<VDC: DX_end>
<VDC: DY_end>

8.5.19 ELLIPTICAL ARC CLOSE

<ELLIPTICAL-ARC-CLOSE-opcode: 3/4 2/7>
<point: centrepoint>
<point: endpoint of first conjugate diameter>
<point: endpoint of second conjugate diameter>
<VDC: DX_start>
<VDC: DY_start>
<VDC: DX_end>
<VDC: DY_end>
<enumerated: close-type>

<enumerated: close-type>   =   <integer: 0>   {pie}
                                          | <integer: 1>   {chord}

8.6 Attribute Elements

8.6.1 LINE BUNDLE INDEX

<LINE-BUNDLE-INDEX-opcode: 3/5 2/0>
<integer: line-bundle-index>

<integer: line-bundle-index>   = <positive integer>

8.6.2 LINE TYPE

<LINE-TYPE-opcode: 3/5 2/1>
<index: line-type>

<index: line-type>   = <integer: 1> {solid}
               | <integer: 2> {dash}
               | <integer: 3> {dot}
               | <integer: 4> {dash-dot}
               | <integer: 5> {dash-dot-dot}
               | <integer: negative> {private line type}

8.6.3 LINE WIDTH

<LINE-WIDTH-opcode: 3/5 2/2>
<line-width-specifier>

<line-width-specifier>   = <real: line width scale factor>
                {if LINE WIDTH SPECIFICATION MODE is scaled}
               | <VDC: line width>
                {if LINE WIDTH SPECIFICATION MODE is absolute}

8.6.4 LINE COLOUR

<LINE-COLOUR-opcode: 3/5 2/3>
<colour-specifier>

<colour-specifier> = <integer: colour-index> {if COLOUR SELECTION MODE is indexed}
         | <RGB> {if COLOUR SELECTION MODE is direct}

<integer: colour-index>   = <non-negative integer>

8.6.5 MARKER BUNDLE INDEX

<MARKER-BUNDLE-INDEX-opcode: 3/5 2/4>
<integer: marker-bundle-index>

<integer: marker-bundle-index> = <positive integer>

8.6.6 MARKER TYPE

<MARKER-TYPE-opcode: 3/5 2/5>
<index: marker-type>

<index: marker-type>   =   <integer: 1>   {dot}
                                      | <integer: 2>   {plus}
                                      | <integer: 3>   {asterisk}
                                      | <integer: 4>   {circle}
                                      | <integer: 5>   {cross (x)}
                                      | <integer: negative>{private marker type}

8.6.7 MARKER SIZE

<MARKER-SIZE-opcode: 3/5 2/6>
<marker-size-specifier>

<marker-size-specifier>   =   <real: marker size scale factor>
                                           {if MARKER SIZE SPECIFICATION MODE is 'scaled'}
                                      | <VDC: marker size>
                                           {if MARKER SIZE SPECIFICATION MODE is 'absolute'}

8.6.8 MARKER COLOUR

<MARKER-COLOUR-opcode: 3/5 2/7>
<colour-specifier: marker-colour>

<colour-specifier>   =   <integer: colour-index>. {if COLOUR SELECTION MODE is 'indexed'}
                        | <RGB> {if COLOUR SELECTION MODE is 'direct'}

<integer: colour-index>   =   <non-negative integer>

8.6.9 TEXT BUNDLE INDEX

<TEXT-BUNDLE-INDEX-opcode: 3/5 3/0>
<integer: text-bundle-index>

<integer: text-bundle-index>   =   <positive integer>

8.6.10 TEXT FONT INDEX

<TEXT-FONT-INDEX-opcode: 3/5 3/1>
<integer: font-index>

<integer: text-font-index>   =   <positive integer>

8.6.11 TEXT PRECISION

<TEXT-PRECISION-opcode: 3/5 3/2>
<enumerated: TEXT-PRECISION>

<enumerated: text-precision>   =   <integer: 0> {string}
                                  ¦   <integer: 1> {character}
                                  ¦   <integer: 2> {stroke}

8.6.12 CHARACTER EXPANSION FACTOR

<CHARACTER-EXPANSION-FACTOR-opcode: 3/5 3/3>
<real: expansion-factor>

<real: expansion-factor>   =   <non-negative real>

8.6.13 CHARACTER SPACING

<CHARACTER-SPACING-opcode: 3/5 3/4>
<real: character-spacing>

8.6.14 TEXT COLOUR

<TEXT-COLOUR-opcode: 3/5 3/5>
<colour-specifier>

<colour-specifier>   =   <integer: colour-index> {if COLOUR SELECTION MODE is 'indexed'}
                        ¦   <RGB> {if COLOUR SELECTION MODE is 'direct'}

<integer: colour-index>   =   <non-negative integer>

8.6.15 CHARACTER HEIGHT

<CHARACTER-HEIGHT-opcode: 3/5 3/6>
<VDC: character-height>

<VDC: character-height>   =   <non-negative VDC>

8.6.16 CHARACTER ORIENTATION

<CHARACTER-ORIENTATION-opcode: 3/5 3/7>
<VDC: x-component of up vector>
<VDC: y-component of up vector>
<VDC: x-component of base vector>
<VDC: y-component of base vector>

8.6.17 TEXT PATH

<TEXT-PATH-opcode: 3/5 3/8>
<enumerated: text-path>

<enumerated: text-path>   =   <integer: 0> {right}
                                  ¦   <integer: 1> {left}
                                  ¦   <integer: 2> {up}

     | <integer: 3> {down}

8.6.18 TEXT ALIGNMENT

<TEXT-ALIGNMENT-opcode: 3/5 3/9>
<enumerated: horizontal-alignment>
<enumerated: vertical-alignment>
<real: continuous-horizontal-alignment>
<real: continuous-vertical-alignment>

<enumerated: horizontal-alignment> = <integer: 0> {normal}
                                        | <integer: 1> {left}
                                        | <integer: 2> {centre}
                                        | <integer: 3> {right}
                                        | <integer: 4> {continuous horizontal}

<enumerated: vertical-alignment>   = <integer: 0> {normal}
                                        | <integer: 1> {top}
                                        | <integer: 2> {cap}
                                        | <integer: 3> {half}
                                        | <integer: 4> {base}
                                        | <integer: 5> {bottom}
                                        | <integer: 6> {continuous vertical}

8.6.19 CHARACTER SET INDEX

<CHARACTER-SET-INDEX-opcode: 3/5 3/10>
<integer: character-set-index>

<integer: character-set-index>   =   <positive integer>

8.6.20 ALTERNATE CHARACTER SET INDEX

<ALTERNATE-CHARACTER-SET-INDEX-opcode: 3/5 3/11>
<integer: alternate-character-set-index>

<integer: alternate-character-set-index> =<positive integer>

8.6.21 FILL BUNDLE INDEX

<FILL-BUNDLE-INDEX-opcode: 3/6 2/0>
<integer: fill-bundle-index>

<integer: fill-bundle-index>   =   <positive integer>

8.6.22 INTERIOR STYLE

<INTERIOR-STYLE-opcode: 3/6 2/1>
<enumerated: interior-style>

```
<enumerated: interior-style>    =    <integer: 0>        {hollow}
                                |    <integer: 1>        {solid}
                                |    <integer: 2>        {pattern}
                                |    <integer: 3>        {hatch}
                                |    <integer: 4>        {empty}
                                |    <integer: negative> {private style}
```

8.6.23 FILL COLOUR

```
<FILL-COLOUR-opcode: 3/6 2/2>
<colour-specifier>

<colour-specifier>    =    <integer: colour-index> {if COLOUR SELECTION MODE is indexed}
                      |    <RGB> {if COLOUR SELECTION MODE is direct}

<integer: colour-index>             =    <non-negative integer>
```

8.6.24 HATCH INDEX

```
<HATCH-INDEX-opcode: 3/6 2/3>
<index: hatch-index>

<index: hatch-index>    =    <integer: 1>        {horizontal}
                        |    <integer: 2>        {vertical}
                        |    <integer: 3>        {positive slope}
                        |    <integer: 4>        {negative slope}
                        |    <integer: 5>        {horizontal/vertical cross}
                        |    <integer: 6>        {positive/negative cross}
                        |    <integer: negative> {private styles}
```

8.6.25 PATTERN INDEX

```
<PATTERN-opcode: 3/6 2/4>
<integer: pattern-index>

<integer: pattern-index>            =    <positive integer>
```

8.6.26 EDGE BUNDLE INDEX

```
<EDGE-BUNDLE-INDEX-opcode: 3/6 2/5>
<integer: edge-bundle-index>

<integer: edge-bundle-index>        =    <positive integer>
```

8.6.27 EDGE TYPE

```
<EDGE-TYPE-opcode: 3/6 2/6>
<index: edge-type>
```

<index: edge-type>　　　　　　=　<integer: 1>　　{solid}
　　　　　　　　　　　　　　　｜　<integer: 2>　　{dash}
　　　　　　　　　　　　　　　｜　<integer: 3>　　{dot}
　　　　　　　　　　　　　　　｜　<integer: 4>　　{dash-dot}
　　　　　　　　　　　　　　　｜　<integer: 5>　　{dash-dot-dot}
　　　　　　　　　　　　　　　｜　<integer: negative> {private edge type}

8.6.28 EDGE WIDTH

<EDGE-WIDTH-opcode: 3/6 2/7>
<edge-width-specifier>

<edge-width-specifier>　　　　　=　<real: edge width scale factor>
　　　　　　　　　　　　　　　　　{if EDGE WIDTH SPECIFICATION MODE is 'scaled'}
　　　　　　　　　　　　　　　｜　<VDC: edge width>
　　　　　　　　　　　　　　　　　{if EDGE WIDTH SPECIFICATION MODE is 'absolute'}

8.6.29 EDGE COLOUR

<EDGE-COLOUR-opcode: 3/6 2/8>
<colour-specifier>

<colour-specifier>　　=　<integer: colour-index> {if COLOUR SELECTION MODE is 'indexed'}
　　　　　　　　　　｜　<RGB> {if COLOUR SELECTION MODE is 'direct'}

<integer: colour-index>　　　　　=　<non-negative integer>

8.6.30 EDGE VISIBILITY

<EDGE-VISIBILITY-opcode: 3/6 2/9>
<enumerated: edge visibility>

<enumerated: edge visibility>　　=　<integer: 0>　{off}
　　　　　　　　　　　　　　　｜　<integer: 1>　{on}

8.6.31 FILL REFERENCE POINT

<FILL-REFERENCE-POINT-opcode: 3/6 2/10>
<point: fill-reference-point>

8.6.32 PATTERN TABLE

<PATTERN-TABLE-opcode: 3/6 2/11>
<integer: pattern-table-index>
<integer: nx>
<integer: ny>
<integer: local-colour-precision> {see 8.5.9, CELL ARRAY, for elaboration}
<colour-list> {see 8.5.9, CELL ARRAY, for elaboration}

33 PATTERN SIZE

PATTERN-SIZE-opcode: 3/6 2/12>
 VDC: pattern-height-vector-x-component>
 VDC: pattern-height-vector-y-component>
 <VDC: pattern-width-vector-x-component>
 <VDC: pattern-width-vector-y-component>

.34 COLOUR TABLE

<COLOUR-TABLE-opcode: 3/6 3/0>
 <integer: starting-index>
 <direct-colour-list>

.35 ASPECT SOURCE FLAGS

<ASPECT-SOURCE-FLAGS-opcode: 3/6 3/1>
 <aspect-pair>+

<aspect-pair>   =   <aspect> <aspect-source>

<aspect>   =   <integer: 0> {line type asf}
                                          <integer: 1> {line width asf}
                                          <integer: 2> {line colour asf}
                                          <integer: 3> {marker type asf}
                                          <integer: 4> {marker size asf}
                                          <integer: 5> {marker colour asf}
                                          <integer: 6> {text font index asf}
                                          <integer: 7> {text precision asf}
                                          <integer: 8> {character expansion factor asf}
                                          <integer: 9> {character spacing asf}
                                          <integer: 10> {text colour asf}
                                          <integer: 11> {interior style asf}
                                          <integer: 12> {fill colour asf}
                                          <integer: 13> {hatch index asf}
                                          <integer: 14> {pattern index asf}
                                          <integer: 15> {edge type asf}
                                          <integer: 16> {edge width asf}
                                          <integer: 17> {edge colour asf}
                                          <integer: 511> {all}
                                          <integer: 510> {all line}
                                          <integer: 509> {all marker}
                                          <integer: 508> {all text}
                                          <integer: 507> {all fill}
                                          <integer: 506> {all edge}

In addition to encoding the 18 ASFs associated with attributes that may be bundled, this encoding
defines 6 pseudo-asfs:

all: set all ASFs as indicated;

all line: set line type, line width, and line colour ASFs as indicated.

all marker: set marker type, marker size, and marker colour ASFs as indicated.

all text: set text font index, text precision, character expansion factor, character spacing,
                and text colour ASFs as indicated.

all fill: set interior style, hatch index, pattern index, and fill colour ASFs as indicated.

all edge: set edge type, edge width, and edge colour ASFs as indicated.

<aspect-source>   =   <integer:0> {individual}
                                        | <integer:1> {bundled}

8.7 Escape Elements

8.7.1 ESCAPE

<ESCAPE-opcode: 3/7 2/0>
<integer: identifier>
<data-record: opcode-and-parameters-of-private-command>

8.7.2 DOMAIN RING

<DOMAIN-RING-opcode: 3/7 3/0>
<enumerated: angular-resolution-factor>
<integer: radius>

<enumerated: angular-resolution-factor>   =<integer: 0>
                                           ¦   <integer: 1>
                                           ¦   <integer: 2>
                                           ¦   <integer: 3>

This primitive specifies the precision of the coordinate encoding, when using incremental mode.

The actual number of points on a ring with a given ring size is determined by the angular-resolution-factor. If the ring size parameter value is zero, the basic ring size is used, which is 1 for VDC TYPE integer. For VDC TYPE real the basic ring size is determined from the current smallest-real-code (as set with VDC REAL PRECISION).

8.8 External Elements

8.8.1 MESSAGE

<MESSAGE-opcode: 3/7 2/1>
<enumerated: action-flag>
<string: message-to-be-displayed>

<enumerated: flag>   =   <integer: 0> {no action required}
                                       ¦   <integer: 1> {action required}

8.8.2 APPLICATION DATA

<APPLICATION-DATA-opcode: 3/7 2/2>
<integer: identifier>
<data-record: application-data>

9 Defaults

Parameter precisions for the character-coded graphics encoding:

REAL PRECISION:
    largest-real-code — 10
    smallest-real-code — -10
    default-exponent-for-reals — -10
    exponents-allowed — 1 (forbidden)

INTEGER PRECISION: 10

INDEX PRECISION: 10

COLOUR PRECISION: 6

COLOUR INDEX PRECISION: 10

COLOUR VALUE EXTENT:
    minimum-colour-value 0,0,0
    maximum-colour-value 63,63,63

VDC REAL PRECISION:
    largest-real-code — 10
    smallest-real-code — -10
    default-exponent-for-reals — -10
    exponents-allowed — 1 (forbidden)

VDC INTEGER PRECISION: 20

DOMAIN RING
    angular resolution factor — 0;
    ring size for VDC type integer — 1;
    ring size for VDC type real if smallest-real code $(src) < -8 — 2^{(-8-src)}$;
    ring size for VDC type real if smallest-real code $(src) \geq -8 — 1$.

10 Conformance

A metafile conforms to this character-coded encoding if it meets the following requirements:

— Each metafile element described in this part is coded in the manner described.

— Private (non-standard) metafile elements are all coded using the ESCAPE and GDP metafile elements. Opcodes reserved for future standardization are not used to code private (non-standard) metafile elements.

— Private (non-standard) values of index and enumerated parameters are all coded using negative integers. In coding index parameters, a metafile shall not use non-negative integers to represent private values of index parameters.

A conforming metafile may include, within the string parameters of TEXT, RESTRICTED TEXT, and APPEND TEXT elements, as well as string parameters within the data records of GENERALIZED DRAWING PRIMITIVE (GDP) elements (their admissibility here depends upon the particular GDP definition), the ISO 2022 controls for designating and invoking G-sets, but shall not include the escape sequence used to enter the encoding from ISO 2022, except as the first character of the file.

Part 2

ANNEXES

A Formal Grammar

In this encoding scheme, a metafile element always has the form

```
<metafile element>        ::=  <opcode>
                               <parameter character>*
```

A CGM opcode is encoded as a single-byte, double-byte, or multi-byte character. Single-byte characters are used for Graphical Primitive Elements, except for some of the Circular and Elliptical Output Elements, where double-byte characters have been assigned. Multi-byte opcodes are used to extend the available opcode space. For all other metafile elements, two or more CGM elements share the same primary opcode character, and a secondary opcode indicates the particular element.

The enumerated types are represented by integers. The specific integer values for each enumerated type are listed in clause 8 in the format

```
  <integer: 0>   {first enumerated element}
¦ <integer: 1>   {second enumerated element}
¦ <integer: 2>   {third enumerated element}
```

Negative integers denote private values of the enumerated types.

The other terminal symbols are described in detail in clause 8. A reference to the relevant sections is given here.

```
<integer>                 ::=  <more significant bits>*
                               <least significant bits>
```

where the least significant bits contain the sign bit. See 6.3.

```
<real>                    ::=  <integral mantissa>
                               <exponent>(o)

<integral mantissa>       ::=  <integer>

<exponent>                ::=  <integer>
```

where the least significant bits of the integral mantissa indicate whether an exponent follows.

```
<point>                   ::=  <integer> <integer>
                          ¦    <real> <real>
```

where a point list is encoded such that the first point is absolute within VDC space and each of the following points is relative to its previous one. See 6.5.

```
<vdc value>               ::=  <integer>
                          ¦    <real>
```

Formal Grammar

\<string\>	::=	\<SOS\> \<character\>* \<ST\>
\<character\>	::=	a character from the 7-bit or 8-bit code tables, as discussed in 3.1.
\<character substitution\>	::=	\<string\> as described in clause 7.
\<colour index\>	::=	\<integer\>
\<red green blue\>	::=	\<bits for red\> \<bits for green\> \<bits for blue\>

The packing of these bits is described in 6.7.

\<integer prec value\>	::=	\<integer\> See 8.2.
\<real prec value\>	::=	\<integer\> \<integer\> \<integer\> \<integer\> See 8.2
\<index prec value\>	::=	\<integer\> See 8.2.
\<colour prec value\>	::=	\<integer\> See 8.2.
\<colour index prec value\>	::=	\<integer\> See 8.2.
\<vdc integer prec value\>	::=	\<integer\> See 8.4.
\<vdc real prec value\>	::=	\<integer\> \<integer\> \<integer\> \<integer\> See 8.4.
\<colour list\>	::=	\<indexed colour list\> ¦ \<direct colour list\>
\<indexed colour list\>	::=	\<indexed normal colour list\> ¦ \<indexed bitstream colour list\> ¦ \<indexed runlength colour list\> ¦ \<indexed runlength bitstream colour list\>
\<direct colour list\>	::=	\<direct normal colour list\> ¦ \<direct bitstream colour list\> ¦ \<direct runlength colour list\> ¦ \<direct runlength bitstream colour list\>
\<indexed normal colour list\>	::=	\<integer: 0\> \<basic format\>+
\<indexed bitstream colour list\>	::=	\<integer: 1\> \<bitstream: concatenation of colour indexes\>

<indexed runlength colour list> ::= <integer: 2>
                                  <indexed run>+

<indexed run>   ::= <integer: colour index>
                                  <integer: number of repetitions>

<indexed runlength bitstream colour list>::=<integer: 3>
                                  <indexed bitstream run>+

<indexed bitstream run>   ::= <bitstream: colour index>
                                  <integer: number of repetitions>

<direct normal colour list> ::= <integer: 0>
                                  <RGB: colour value>+

<direct bitstream colour list> ::= <integer: 1>
                                  <bitstream: concatenation of colour values>

<direct runlength colour list> ::= <integer: 2>
                                  <direct run>+

<direct run>   ::= <RGB: colour value>
                                  <integer: number of repetitions>

<direct runlength bitstream colour list>::=<integer: 3>
                                  <direct run>+

<escape data record>   ::= <data record> See 8.7.

<GDP data record>   ::= <data record> See 8.5.

<data record>   ::= <data item>+

<data item>   ::= <integer>
                                  ¦ <real>
                                  ¦ <string>
                                  ¦ <bitstream>

ANSI X3.122 - 1986

Information Processing Systems

Computer Graphics

Metafile for the Storage and Transfer
of Picture Description Information

Part 3

Binary Encoding

ANSI X3.122 - 1986
Part 3

CONTENTS

0 Introduction

0.1 Purpose of the Binary Encoding

The Binary Encoding of the Computer Graphics Metafile (CGM) provides a representation of the Metafile syntax that can be optimized for speed of generation and interpretation, while still providing a standard means of interchange among computer systems. The encoding uses binary data formats that are much more similar to the data representations used within computer systems than the data formats of the other encodings.

Some of the data formats may exactly match those of some computer systems. In such cases processing is reduced very much relative to the other standardized encodings. On most computer systems processing requirements for the Binary Encoding will be substantially lower than for the other encodings.

In cases where a computer system's architecture does not match the standard formats used in the Binary Encoding, and where absolute minimization of processing requirements is critical, and where interchange among dissimilar systems does not matter, it may be more appropriate to use a private encoding, conforming to the rules specified in clause 7 of part 1 of this Standard.

0.2 Objectives

This encoding has the following features.

a) Partitioning of parameter lists: metafile elements are coded in the Binary Encoding by one or more partitions (see clause 4); the first (or only) partition of an element contains the opcode (Element Class plus Element Id).

b) Alignment of elements: every element begins on a word boundary. Alignment of partitions that require an odd number of octets is effected by padding with an octet with all bits zero. A no-op element is available in this encoding; it is ignored. It may be used to align data on machine-dependent record boundaries for speed of processing.

c) Uniformity of format: all elements have an associated parameter length value. The length is specified as an octet count. As a result, it is possible to scan the metafile, without interpreting it, at high speed.

d) Alignment of coordinate data: at default precisions and by virtue of alignment of elements, coordinate data always start on word boundaries. This minimizes processing by ensuring, on a wide class of computing systems, that single coordinates do not have to be assembled from pieces of multiple computer words.

e) Efficiency of encoding integer data: other data such as indexes, colour and characters are encoded as one or more octets. The precision of every parameter is determined by the appropriate precision as given in the METAFILE DESCRIPTOR.

f) Order of bit data: in each word, or unit within a word, the bit with the highest number is the most significant bit. Likewise, when data word are accessed sequentially, the least significant word follows the most significant.

g) Extensibility: the arrangement of Element Class and Element Id values has been designed to allow future growth, such as new graphical elements.

h) Format of real data: real numbers are encoded using either IEEE floating point representation or a metafile fixed-point representation.

i) Run length encoding: if many adjacent cells have the same colour (or colour index) efficient encoding is possible. For each run, the colour (or colour index) is specified, followed by a cell count.

j)   Packed list encoding: if adjacent colour cells do not have the same colour (or colour index) the metafile provides bit-stream lists in which the values are packed as closely as possible.

0.3 Relationship to Other Standards

The floating point representation of real data in this part of the Standard is that in ANSI/IEEE 754-1986·

The representation of character data in this part of the Standard follows the rules of ISO 646 and ISO 2022.

For certain elements, the CGM defines value ranges as being reserved for registration. The values and their meanings will be defined using the established procedures (see part 1, 4.11.)

0.4 Status of Annexes

The annexes do not form an integral part of the Standard but are included to provide extra information and explanation.

1 Scope and Field of Application

1.1 Scope

This part of this Standard specifies a binary encoding of the Computer Graphics Metafile. For each of the elements specified in part 1 of this Standard, an encoding is specified in terms of a data type. For each of these data types, an explicit representation in terms of bits, octets and words is specified. For some data types, the exact representation is a function of the precisions being used in the metafile, as recorded in the METAFILE DESCRIPTOR.

1.2 Field of Application

This encoding of the Computer Graphics Metafile will, in many circumstances, minimize the effort required to generate and interpret the metafile.

2 References

ISO 646 Information Processing - 7-bit coded character set for information interchange

ISO 2022 Information Processing - ISO 7-bit and 8-bit coded character sets - Code Extension techniques

ANSI/IEEE 754 Standard for Binary Floating Point Arithmetic

3 Notational Conventions

"Command Header" is used throughout this part to refer to that portion of a Binary-Encoded element that contains the opcode (element class plus element id) and parameter length information (see clause 4).

Within this part, the terms "octet" and "word" have specific meanings. These meanings may not match those of a particular computer system on which this encoding of the metafile is used.

An octet is an 8-bit entity. All bits are significant. The bits are numbered from 7 (most significant) to 0 (least significant).

A word is a 16-bit entity. All bits are significant. The bits are numbered from 15 (most significant) to 0 (least significant).

4 Overall Structure

4.1 General Form of Metafile

All elements in the metafile are encoded using a uniform scheme. The elements are represented as variable length data structures, each consisting of opcode information (element class plus element id) designating the particular element, the length of its parameter data and finally the parameter data (if any).

The structure of the metafile is as follows. (For the purposes of this diagram only, MF is used as an abbreviation for METAFILE.)

| BEGIN MF | MD | <picture> ... | END MF |

The BEGIN METAFILE element is followed by the METAFILE DESCRIPTOR (MD). After this the pictures follow, each logically independent of each other. Finally the Metafile is ended with an END METAFILE element.

4.2 General Form of Pictures

Apart from the BEGIN METAFILE, END METAFILE and Metafile Descriptor elements, the metafile is partitioned into pictures. All pictures are mutually independent. A picture consists of a BEGIN PICTURE element, a PICTURE DESCRIPTOR (PD) element, a BEGIN PICTURE BODY element, an arbitrary number of control, graphical and attribute elements and finally an END PICTURE element. (For the purpose of this diagram only, PIC is used as an abbreviation for PICTURE and BEGIN BODY for BEGIN PICTURE BODY.)

| BEGIN PIC | PD | BEGIN BODY | <element> ... | END PIC |

4.3 General Structure of the Binary Metafile

The binary encoding of the metafile is a logical data structure consisting of a sequential collection of bits. For convenience in describing the length and alignment of metafile elements, fields of two different sizes are defined within the structure. These fields are used in the remainder of this part of the Standard for illustrating the contents and structure of elements and parameters.

For measuring the lengths of elements the metafile is partitioned into octets, which are 8-bit fields.

The structure is also partitioned into 16-bit fields called words (these are logical metafile words). To optimize processing of the binary metafile on a wide collection of computers, metafile elements are constrained to start on word boundaries within the binary data structure (this alignment may necessitate padding an element with bits to a word boundary if the parameter data of the element does not fill to such a boundary).

The octet is the fundamental unit of organization of the binary metafile.

The bits of an octet are numbered 7 to 0, with 7 being the most significant bit. The bits of a word are numbered 15 to 0, with 15 being the most significant bit.

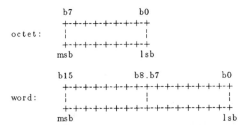

If the consecutive bits of the binary data structure are numbered 1..N, and the consecutive octets are numbered 1..N/8, and the consecutive words are numbered 1..N/16, then the logical correspondence of bits, octets, and words in the binary data structure is as illustrated in the following table:

metafile bit number	octet bit number	word bit number
1	b7/octet1	b15/word1
.	.	.
.	.	.
8	b0/octet1	b8/word1
9	b7/octet2	b7/word1
.	.	.
.	.	.
16	b0/octet2	b0/word1
17	b7/octet3	b15/word2
.	.	.
.	.	.
24	b0/octet3	b8/word2
25	b7/octet4	b7/word2
.	.	.

4.4 Structure of the Command Header

Throughout this sub-clause, the term "command" is used to denote a binary-encoded element. Metafile elements are represented in the Binary Encoding in one of two forms — short-form commands and long-form commands. There are two differences between them:

— a short-form command always contains a complete element; the long-form command can accommodate partial elements (the data lists of elements can be partitioned);

— a short-form command only accommodates parameter lists up to 30 octets in length; the long-form command accommodates lengths up to 32767 octets per data partition.

The forms differ in the format of the Command Header that precedes the parameter list. The command form for an element (short or long) is established by the first word of the element. For the short-form, the Command Header consists of a single word divided into three fields: element class, element id and parameter list length.

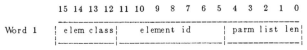

Figure 1. Format of a short-form Command Header.

ANSI X3.122 - 1986 Part 3

The fields in the short-form Command Header are as follows.

　　bits 15-12　element class (value range 0-15)

　　bits 11-5　element id (value range 0-127)

　　bits 4-0　parameter list length: the number of octets of parameter data that follow for this command (value range 0-30)

This Command Header is then followed by the parameter list.

The first word of a long-form command is identical in structure to the first word of a short-form command. The presence of the value 11111 binary (decimal 31) in parameter list length field indicates that the command is a long-form command. The Command Header for the long-form command consists of two words. The second word contains the actual parameter list length. The two header words are then followed by the parameter list.

In addition to allowing longer parameter lists, the long-form command allows the parameter list to be partitioned. Bit 15 of the second word indicates whether the given data complete the element or more data follow. For subsequent data partitions of the element, the first word of the long-form Command Header (containing element class and element id) is omitted; only the second word, containing the parameter list length, is given. The parameter list length for each partition specifies the length of that partition, not the length of the complete element. The final partition of an element is indicated by bit 15 of the parameter list length word being zero.

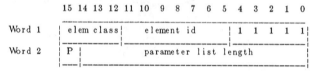

Figure 2. Format of a long-form Command Header.

The fields in the long-form Command Header are as follows.

Word 1

　　bits 15-12　element class (value range 0-15)

　　bits 11-5　element id (value range 0-127)

　　bits 4-0　binary value 11111 (decimal 31) indicating long-form

Word 2

　　bit 15　partition flag

　　　　— 0 for 'last' partition

　　　　— 1 for 'not-last' partition

　　bits 14-0　parameter list length: the number of octets of parameter data that follow for this command or partition (value range 0-32767).

The parameter values follow the parameter list length for either the long-form or short-form commands. The number of values is determined from the parameter list length and the type and precision of the operands. These parameter values have the format illustrated in clause 5 of this part. The parameter type for coordinates is indicated in the Metafile Descriptor. For non-coordinate parameters, the parameter type is as specified in clause 5 of part 1. If the parameter type is encoding dependent, its code is specified in the coding tables of clause 7 of this part. Unless otherwise stated, the order of parameters is as listed in clause 5 of part 1.

Every command is constrained to begin on a word boundary. This necessitates padding the command with a single null octet at the end of the command if the command contains an odd number of octets of parameter data. In addition, in elements with parameters whose precisions are shorter than one octet (i.e., those containing a 'local colour precision' parameter) it is necessary to pad the last data-containing octet with null bits if the data do not fill the octet. In all cases, the parameter list length is the count of octets actually containing parameter data — it does not include the padding octet if one is present. It is only at the end of a command that padding is performed, with the single exception of the CELL ARRAY element.

The purpose of this command alignment constraint is to optimize processing on a wide class of computers. At the default metafile precisions, the parameters which are expected to occur in greatest numbers (coordinates, etc) will align on 16-bit boundaries, and Command Headers will align on 16-bit boundaries. Thus, at the default precisions the most frequently parsed entities will lie entirely within machine words in a large number of computer designs. The avoidance of assembling single metafile parameters from pieces of several computer words will approximately halve the amount of processing required to recover element parameters and command header fields from a binary metafile data stream.

This optimization may be compromised or destroyed altogether if the metafile precisions are changed from default. Commands are still constrained to begin on 16-bit boundaries, but the most frequently expected parameters may no longer align on such boundaries as they do at the default precisions.

The short form command header with element class 15, element id 127, and parameter list length 0 is reserved for extension of the number of available element classes in future revisions of this Standard. It should be treated by interpreters as any other element, as far as parsing is concerned. The next "normal" element encountered will have an actual class value different from that encountered in the "element class" field of the command header — it will be adjusted by a bias as will be defined in a future revision of this Standard.

5 Primitive Data Forms

The Binary Encoding of the CGM uses five primitive data forms to represent the various abstract data types used to describe parameters in the part 1 of this Standard.

The primitive data forms and the symbols used to represent them are as follows.

SI   Signed Integer
UI   Unsigned Integer
C   Character
FX   Fixed Point Real
FP   Floating Point Real

Each of these primitive forms (except Character) can be used in a number of precisions. The definitions of the primitive data forms in sub-clauses 5.1 to 5.5 show the allowed precisions for each primitive data form. The definitions are in terms of 'metafile words' which are 16-bit units.

The following terms are used in the following diagrams when displaying the form of numeric values.

msb   most significant bit
lsb   least significant bit
S   sign bit

The data types in the following data diagrams are illustrated for the case that the parameter begins on a metafile word boundary. In general, parameters may align on odd or even octet boundaries, because they may be preceded by an odd or even number of octets of other parameter data. Elements containing the local colour precision parameter may have parameters shorter than one octet. It is possible in such cases that the parameters will not align on octet boundaries.

5.1 Signed Integer

Signed integers are represented in "two's complement" format. Four precisions may be specified for signed integers: 8-bit, 16-bit, 24-bit and 32-bit. (Integer coordinate data encoded with this primitive data form do not use the 8-bit precision.)

5.1.1 Signed Integer at 8-bit precision

Each value occupies half a metafile word (one octet).

```
   15 14 13 12 11 10  9  8  7  6  5  4  3  2  1  0
  _____
 |S|msb     value i    lsb|S|msb    value i+1  lsb|
 |-|----------------------|-|----------------------|
```

5.1.2 Signed Integer at 16-bit precision

Each value occupies one metafile word.

```
   15 14 13 12 11 10  9  8  7  6  5  4  3  2  1  0
  _____
 |  S|msb                value               lsb|
 |--|------------------------------------------------|
```

5.1.3 Signed Integer at 24-bit precision

Each value straddles two successive metafile words.

```
        15 14 13 12 11 10  9  8  7  6  5  4  3  2  1  0
       _____
Word 1 |  S¦msb              value i
       |__¦_____
Word 2 |        value i        lsb¦S¦msb     value i+1
       _____¦_¦_____
Word 3                       value i+1                lsb¦
       _____|
```

5.1.4 Signed Integer at 32-bit precision

Each value fills two complete metafile words.

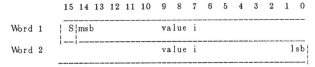

```
        15 14 13 12 11 10  9  8  7  6  5  4  3  2  1  0
       _____
Word 1 |  S¦msb              value i
       |__¦_____
Word 2 |               value i                      lsb¦
       _____|
```

5.2 Unsigned Integer

Four precisions may be specified for unsigned integers: 8-bit, 16-bit, 24-bit and 32-bit.

5.2.1 Unsigned Integers at 8-bit precision

Each value occupies half a metafile word.

```
        15 14 13 12 11 10  9  8  7  6  5  4  3  2  1  0
       _____
       ¦msb     value i        lsb¦msb     value i+1   lsb¦
       |_____¦_____|
```

5.2.2 Unsigned Integers at 16-bit precision

Each value occupies one metafile word.

```
        15 14 13 12 11 10  9  8  7  6  5  4  3  2  1  0
       _____
       ¦msb                    value                  lsb¦
       |_____|
```

5.2.3 Unsigned Integers at 24-bit precision

Each value straddles two successive metafile words.

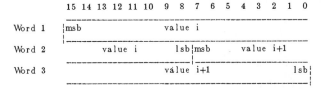

```
        15 14 13 12 11 10  9  8  7  6  5  4  3  2  1  0
       _____
Word 1 ¦msb                  value i
       ¦_____
Word 2 |        value i        lsb¦msb     value i+1
       _____¦_____
Word 3                       value i+1                lsb¦
       _____|
```

5.2.4 Unsigned Integers at 32-bit precision

Each value fills two complete metafile words.

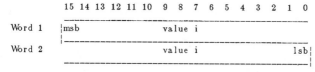

```
        15 14 13 12 11 10  9  8  7  6  5  4  3  2  1  0
       _____
Word 1 |msb                value i
       |_____
Word 2                     value i                  lsb|
       _____|
```

5.3 Character

Each character is stored in an octet.

```
        15 14 13 12 11 10  9  8  7  6  5  4  3  2  1  0
       _____
       |      Character i      |    Character i+1      |
       |_____ |_____|
```

5.4 Fixed Point Real

Fixed point real values are stored as two integers; the first represents the "whole part" and has the same form as a Signed Integer (SI; see 5.1); the second represents the "fractional part" and has the same form as an Unsigned Integer (UI; see sub-clause 5.2). Two precisions may be specified for Fixed Point Reals: 32-bit or 64-bit.

5.4.1 Fixed Point Real at 32-bit precision

Each Fixed Point Real occupies 2 complete metafile words; the first has the form of a 16-bit Signed Integer and the second the form of a 16-bit Unsigned Integer.

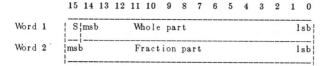

```
        15 14 13 12 11 10  9  8  7  6  5  4  3  2  1  0
       _____
Word 1 | S|msb       Whole part                   lsb|
       |__|_____
Word 2 |msb          Fraction part                 lsb|
       |_____|
```

5.4.2 Fixed Point Real at 64-bit precision

Each Fixed Point Real occupies 4 complete metafile words; the first has the form of a 32-bit Signed Integer and the second the form of a 32-bit Unsigned Integer.

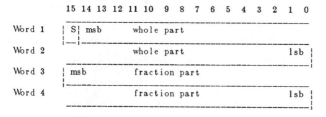

```
        15 14 13 12 11 10  9  8  7  6  5  4  3  2  1  0
       _____
Word 1 | S|  msb       whole part
       |__|_____
Word 2               whole part                    lsb |
       _____|
Word 3 |  msb        fraction part
       |_____
Word 4               fraction part                 lsb |
       _____|
```

5.4.3 Value of Fixed Point Reals

The values of the represented real numbers are given by:

for 32 bits: $\text{real\_value} = SI + \left(\dfrac{UI}{2^{16}} \right)$

for 64 bits: $\text{real\_value} = SI + \left(\dfrac{UI}{2^{32}} \right)$

SI stands for the "whole part" and UI stands for the "fractional part" in these equations. SI, the whole part, is the largest integer less than or equal to the real number being represented.

5.5 Floating Point

Floating Point Real values are represented in the floating point format of IEEE standard P754. This format contains three parts:

— a sign bit ('s');

— a biased exponent part ('e');

— a fraction part ('f').

The value is a function of these three values ('s', 'e' and 'f'). If 's' is '0', the value is positive; if 's' is '1', the value is negative. Two precisions may be specified for Floating Point Reals: 32-bit or 64-bit. The magnitude of the value is calculated as follows for 32-bit representation.

a) If e = 255 and f $\neq$ 0, then the value is undefined.

b) If e = 255 and f = 0, then the value is as large a positive (s=0) or negative (s=1) value as possible.

c) If $0 < e < 255$, then the magnitude of the value is $(1.f)(2^{e-127})$.

d) If e = 0 and f $\neq$ 0, then the magnitude of the value is $(0.f)(2^{-126})$.

e) If e = 0 and f = 0, then the value is 0.

The magnitude of the value is calculated as follows for 64-bit representation.

a) If e = 2047 and f $\neq$ 0, then the value is undefined.

b) If e = 2047 and f = 0, then the value is as large a positive (s=0) or negative (s=1) value as possible.

c) If $0 < e < 2047$, then the magnitude of the value is $(1.f)(2^{e-1023})$.

d) If e = 0 and f $\neq$ 0, then the magnitude of the value is $(0.f)(2^{-1022})$.

e) If e = 0 and f = 0, then the value is 0.

5.5.1 Floating Point Real at 32-bit precision

Each Floating Point Real value occupies 2 metafile words. The size of each field in the value is as follows.

sign	1 bit
exponent	8 bits
fraction	23 bits

```
       15 14 13 12 11 10  9  8  7  6  5  4  3  2  1  0
       ------------------------------------------------
Word 1  | S |msb     Exponent    1sb |msb     Fraction
        |-- |--------------------------|----------------
Word 2             Fraction                        1sb |
       ------------------------------------------------
```

5.5.2 Floating Point Real at 64-bit precision

Each Floating Point Real value occupies 4 metafile words. The size of each field in the value is as follows.

sign	1 bit
exponent	11 bits
fraction	52 bits

```
       15 14 13 12 11 10  9  8  7  6  5  4  3  2  1  0
       ------------------------------------------------
Word 1  | S |msb        Exponent      1sb |msb
        |-- |--------------------------------|----------
Word 2                 Fraction
       ------------------------------------------------
Word 3                 Fraction
       ------------------------------------------------
Word 4                 Fraction                    1sb |
       ------------------------------------------------
```

6 Representation of Abstract Parameter Types

The table 1 shows, for each of the abstract parameter types, how it is represented in the Binary Encoding of the CGM in terms of primitive data forms. The columns of the table are as follows.

1) The symbol for the abstract parameter type, as it is specified in clause 5 of part 1 of this Standard.

2) The way the parameter type is constructed in terms of the primitive data forms, at the appropriate precisions. The precisions are those defined in clause 5 of part 1 of this Standard.

3) The symbol for the number of octets required to represent one instance (occurance) of the given parameter, at the given precision, and the formula for computing the number.

4) The symbol for the range of values which the parameter can assume, followed by the numerical values which the parameter can assume, followed by the numerical values which define the range.

The symbols of columns 3 and 4 are used extensively in the code tables in clause 7. Also used in the code tables are variations on those symbols:

+IR, +RR, ..   denote the range of positive integers, range of positive reals, ..

++IR, ++RR, ..   denote the range of non-negative integers, range of non-negative reals, ..

mI, mR ..   denotes 'm' integers, reals, ..

I*, R* ..   denotes an unbounded number of integers, reals, ..

Combinations are used:

2R, 2I, IX*   indicates a parameter that is represented by 2 reals, then a parameter that is represented by 2 integers and finally a parameter that contains an unlimited number of index values.

TABLE 1. Representation of Abstract Data Types

Abstract Symbol	Parameter Construction from Primitive Forms	Octets per Parameter: Symbol and Value	Parameter Range: Symbol and Value
CI	UI at colour index precision (cip)	BCI {=cip/8}	CIR {0..(2**cip-1)}
CD	3UI at direct colour precision (dcp)	BCD {=3*dcp/8} {see notes 1,2}	CDR {0..(2**dcp-1)}
IX	SI at index precision (ixp)	BIX {=ixp/8}	IXR {-2**(ixp-1) to 2**(ixp-1)-1 }
E	SI at fixed precision (16-bit) {see note 3}	BE {=2}	IXR {-2**(15) to 2**(15)-1 }
I	SI at integer precision (ip)	BI {=ip/8}	IR {-2**(ip-1) to 2**(ip-1)-1 }
R	FP or FX at real precision (rp)	BFP {=sum(rp)/8} {see note 4}	RR {=FPR or FXR, see notes 5,10}
S,D	UI,nC	BS {=n+1 or n+3}	SR {see notes 6,12}
VDC	SI at VDC integer precision (vip) or FP or FX at VDC real precision (vrp)	BVDC {=vip/8} or BVDC {=sum(vrp)/8} {see note 4}	VDCR {-2**(vip-1) to 2**(vip-1)-1 } or VDCR {see notes 1,5,7,8}
P	(VDC,VDC)	BP {=2*BVDC}	VDCR {see notes 1,5,7,8}
CO	CI or CD	BCO {=BCI} or BCO {=BCD}	COR {=CIR} {see notes 9,11} or COR {=CDR}

Notes on table 1

1) For parameters that are composed of multiple identical components (e.g., DIRECT COLOUR, CD, and POINT, P) the range value represents the range of a single component.

2) Direct colour is abstractly a real in the range {0,1}. This is normalized onto the unsigned range CDR in the table.

3) Abstract parameter type Enumeration, E, is encoded identically to abstract type Index, IX, at 16-bit precision.

4) The REAL PRECISION element contains an indicator (fixed or floating point) and two precision components. The symbol "sum(rp)" in the table indicates the sum of the number of bits specified in the two components. The same considerations apply to the VDC REAL PRECISION element and the symbol "sum(vrp)" in the tables. The VDC REAL PRECISION control element may cause 'vrp' to be updated in the body of metafile.

5) FPR and VDCR (when VDC are floating point reals) are computed following the IEEE standard (see clause 5 on the Floating Point Data Form).

6) The range for parameter type S is not applicable. The range for character data is not applicable. A string is encoded as a count (unsigned integer) followed by characters. The encoding of the count is similar to the encoding of length information for metafile commands themselves. If the first octet is in the range 0..254, then it represents the character count for the complete string. If the first octet is 255, then the next 16 bits contain the character count and a continuation flag. The first bit is used as a continuation flag (allowing strings longer than 32767 characters) and the next 15 bits represent the count, 0..32767, for the partial string. If the first bit is 0, then this partial string completes the string parameter. If 1, then this partial string will be followed by another. See CHARACTER SET LIST element.

7) The abstract parameter type VDC, a single VDC value, is either a real or an integer, depending on the declaration of the metafile descriptor function VDC TYPE. Subsequent tables use a single set of symbols, VDC, BVDC and VDCR, recognizing that they are computed differently depending on VDC TYPE.

8) The abstract parameter type VDC is a single value; a point, P, is an ordered pair of VDC.

9) The parameter type symbol CO does not correspond to an abstract parameter type as used in part 1 of this Standard. Rather, it is a convenient shorthand for 'colour', which is either direct colour (CD) or indexed colour (CI), depending on the value specified in the COLOUR SELECTION MODE element. The associated octets per parameter and range symbols, BCO and COR, are thus either BCI and CIR or BCD and CDR respectively depending upon COLOUR SELECTION MODE.

10) To eliminate the need to support IEEE floating point in applications that do not need the dynamic range for parameters of type R and VDC, a fixed point real format is provided for scalars (such as line width, character spacing) and VDC. Fixed point reals consist of a (SI,UI) pair.

Fixed point reals (FX) apply to VDC, and to R parameters for the following elements:

   a)   line width;

   b)   edge width;

   c)   character spacing;

   d)   character expansion factor;

   e)   marker size;

   f)   vertical continuous text alignment;

   g)   horizontal continuous text alignment.

11) CELL ARRAY colour can optionally specify 1, 2, 4, 8, 16, 24 or 32 bit precisions for cell colours, as well as using the default CI or CD precision.

The way in which the colour values in CELL ARRAY is represented is an extension of the representation of single colour values. The CELL ARRAY element has a 'cell representation flag' which may take one of two values:

   0   run length representation
   1   packed representation

For PACKED mode, each row of the cell array is represented by an array of colour values without compression. Each row starts on a word boundary. No row length information is stored since all rows are the same length.

The colour data thus occupies $2n_y(1 + [pn_x/16])$ octets, where n_x is the number of cells per row, n_y is the number of rows, p is the number of bits per colour, and [..] denotes "the greatest integer in .."

For RUN LENGTH encoding, the data for each row begins on a word boundary and consists of run-length-lists for runs of constant colour value. Each 'run-length-list' consists of a count of a number of consecutive cells and the representation of that colour. In terms of the abstract terms above, the colour list is of format <I,CO>* and its length is <BI,BCO>*.

12) Abstract parameter type Data Record, D, is encoded in this part as string data, S. The coding tables in clause 7 will use the symbol D for the parameter type, and will use the S-related symbols for other data about the parameter.

7 Representation of Each Element

7.1 Method of Presentation

The elements are grouped according to their class; there are eight classes.

TABLE 2. List of Element Class Codes

Class	Type of Elements
0	Delimiter elements
1	Metafile Descriptor elements
2	Picture Descriptor elements
3	Control elements
4	Graphical Primitive elements
5	Attribute elements
6	Escape element
7	External elements
8-15	Reserved for future standardization

A complete list of element id codes and element class codes is given in annex C.

For each class this clause contains a sub-clause which consists of a table and a set of notes. The table specifies the metafile element, element id, parameter type, parameter list length, parameter range and default values. The parameter list length is given in octets, which in some cases is constant and in other cases is variable. Any element that contains an odd number of octets is padded with one octet of all zero bits. The parameter list length does not include this extra octet (see 4.4). Following each table there is a set of notes that provide additional details on the coding of each element.

7.2 Delimiter Elements

TABLE 3. Encoding of Delimiter Elements

Element Class 0	Element Id	Parameter Type	Parameter List Length	Parameter Range	Default
no-op	0	see below	n	n/a	n/a
BEGIN METAFILE	1	S	BS	SR	null
END METAFILE	2	n/a	0	n/a	n/a
BEGIN PICTURE	3	S	BS	SR	null
BEGIN PICTURE BODY	4	n/a	0	n/a	n/a
END PICTURE	5	n/a	0	n/a	n/a

NOTES (on table 3)

Code Notes

0   no-op: has 1 parameter:

P1: an arbitrary sequence of n octets, n=0,1,2..

The parameter, unlike all other parameters in the binary encoding, is not constructed from the primitive data forms — it is an arbitrary sequence of zero or more octets for padding purposes.

1   BEGIN METAFILE: has 1 parameter:

P1: (string) metafile name

2   END METAFILE: has no parameters.

3   BEGIN PICTURE: has 1 parameter:

P1: (string) picture name

4   BEGIN PICTURE BODY: has no parameters.

5   END PICTURE: has no parameters.

7.3 Metafile Descriptor Elements

TABLE 4. Encoding of Metafile Descriptor Elements

Element Class 1	Element Id	Parameter Type	Parameter List Length	Parameter Range	Default
METAFILE VERSION	1	I	BI	+IR(1..n)	see below
METAFILE DESCRIPTION	2	S	BS	SR	null
VDC TYPE	3	E	BE	0,1	0
INTEGER PRECISION	4	I	BI	8,16,24,32	16
REAL PRECISION	5	3I	3BI	{0,1}, {9,12,16,32}, {23,52,16,32}	1,16,16
INDEX PRECISION	6	I	BI	8,16,24,32	16
COLOUR PRECISION	7	I	BI	8,16,24,32	8
COLOUR INDEX PRECISION	8	I	BI	8,16,24,32	8
MAXIMUM COLOUR INDEX	9	CI	BCI	CIR	63
COLOUR VALUE EXTENT	10	2CD	2BCD	CDR	see below
METAFILE ELEMENT LIST	11	I,2nIX	BI,2nBIX	++IR,IXR	n/a
METAFILE DEFAULTS REPLACEMENT	12	Metafile elements	variable	Metafile elements	none
FONT LIST	13	nS	nBS	SR	see below
CHARACTER SET LIST	14	n(E,S)	n(BE+BS)	{0..4},SR	see below
CHARACTER CODING ANNOUNCER	15	E	BE	0,1,2,3	0

NOTES (on table 4)

Code Notes

1   METAFILE VERSION: has 1 parameter:

   P1: (integer) metafile version number; for this metafile, always 1

2   METAFILE DESCRIPTION: has 1 parameter:

   P1: (string) metafile descriptive string

3   VDC TYPE: has 1 parameter:

   P1: (enumerated) VDC TYPE: valid values are:

   0   VDC values specified in integers
   1   VDC values specified in reals

4   INTEGER PRECISION: has 1 parameter:

   P1: (integer) integer precision: 8, 16, 24 or 32 are the only valid values

5   REAL PRECISION: has 3 parameters:

P1: (enumerated) form of representation for real values: valid values are:

    0   floating point format
    1   fixed point format

P2: (integer) field width for exponent or whole part (including 1 bit for sign)
P3: (integer) field width for fraction or fractional part

Legal combinations of values are:

P1	P2	P3	Result
0	9	23	32-bit floating point
0	12	52	64-bit floating point
1	16	16	32-bit fixed point
1	32	32	64-bit fixed point

6   INDEX PRECISION: has 1 parameter:

P1: (integer) Index precision: valid values are 8,16,24,32

7   COLOUR PRECISION: has 1 parameter:

P1: (integer) Colour precision: valid values are 8,16,24,32

8   COLOUR INDEX PRECISION: has 1 parameter:

P1: (integer) Colour index precision: valid values are 8,16,24,32

9   MAXIMUM COLOUR INDEX: has 1 parameter:

P1: (colour index) maximum colour index that may be encountered in the metafile.

10   COLOUR VALUE EXTENT: has 2 parameters:

P1: (direct colour value) minimum colour value
P2: (direct colour value) maximum colour value

The defaults for the two parameters are (0,0,0) and (255,255,255), respectively.

11   METAFILE ELEMENTS LIST: has 2 parameters:

P1: (integer) number of elements specified
P2: (index-pair array) List of metafile elements in this metafile. Each element is represented by two values: the first is its element class code (as in table 2) and the second is its element id code (as in tables 3 to 10). These codes are listed in annex C. The shorthand pseudo-elements are represented by:

    drawing set: (-1,0)
    drawing-plus-control set: (-1,1)

12   METAFILE DEFAULTS REPLACEMENT: has 1 parameter that itself contains metafile elements. The structure and format is identical to appropriate metafile element(s).

13   FONT LIST: has a variable parameter list:

P1-Pn: n font names (strings)

The default font list consists of one entry: any font capable of representing the standard national character set based on ISO 646.

14   CHARACTER SET LIST: has a variable number of parameter pairs; for each of these:

P1: (enumerated) CHARACTER SET TYPE: valid codes are:

0   94-character G-set
1   96-character G-set
2   94-character multibyte G-set
3   96-character multibyte G-set
4   complete code
negative for private use

P2: (string) Designation sequence tail; see part 1, 5.3.13.

The default character set list consists of one entry: the value of the enumerated parameter is 0, and the value of the string parameter is the designation sequence tail that is registered for the standard national character set based on ISO 646.

15   CHARACTER CODING ANNOUNCER: has 1 parameter:

P1: (enumerated) character coding announcer: valid values are:

0   basic 7-bit
1   basic 8-bit
2   extended 7-bit
3   extended 8-bit
negative for private use

7.4 Picture Descriptor Element

TABLE 5. Encoding of Picture Descriptor Element

Element Class 2	Element Id	Parameter Type	Parameter List Length	Parameter Range	Default
SCALING MODE	1	E,R (FP)	BE+BFP	{0,1},FPR	0,-
COLOUR SELECTION MODE	2	E	BE	{0,1}	0
LINE WIDTH SPECIFICATION MODE	3	E	BE	{0,1}	1
MARKER SIZE SPECIFICATION MODE	4	E	BE	{0,1}	1
EDGE WIDTH SPECIFICATION MODE	5	E	BE	{0,1}	1
VDC EXTENT	6	2P	2BP	VDCR	see below
BACKGROUND COLOUR	7	CD	BCD	CDR	see below

NOTES (on table 5)

Code   Notes

1   SCALING MODE: has 2 parameters:

     P1: (enumerated) scaling mode: valid values are:

        0   abstract scaling
        1   metric scaling

     P2: (real) metric scaling factor, ignored if P1=0

2   COLOUR SELECTION MODE: has 1 parameter:

     P1: (enumerated) colour selection mode:

        0   indexed colour mode
        1   direct colour mode

3   LINE WIDTH SPECIFICATION MODE: has 1 parameter:

     P1: (enumerated) line width specification mode: valid values are:

        0   absolute
        1   scaled

4   MARKER SIZE SPECIFICATION MODE: has 1 parameter:

     P1: (enumerated) marker size specification mode: valid values are:

        0   absolute
        1   scaled

5   EDGE WIDTH SPECIFICATION MODE: has 1 parameter:

     P1: (enumerated) edge width specification mode: valid values are:

        0   absolute

     1   scaled

6   VDC EXTENT: has 2 parameters:

P1: (point) first point
P2: (point) second point

If VDC TYPE is REAL, default VDC EXTENT is (0.0,0.0) , (0.9999..,0.9999..).

If VDC TYPE is INTEGER, default VDC EXTENT is (0,0) , (32767,32767).

7   BACKGROUND COLOUR: has 1 parameter:

P1: (direct colour) background colour, red,green,blue 3-tuple.

7.5 Control Elements

TABLE 6. Encoding of Control Elements

Element Class 3	Element Id	Parameter Type	Parameter List Length	Parameter Range	Default
VDC INTEGER PRECISION	1	I	BI	16,24,32	16
VDC REAL PRECISION	2	3I	3BI	{0,1}, {9,12,16,32}, {23,52,16,32}	1,16,16
AUXILIARY COLOUR	3	CO	BCO	COR	see below
TRANSPARENCY	4	E	BE	{0,1}	1
CLIP RECTANGLE	5	2P	2BP	VDCR	VDC EXTENT
CLIP INDICATOR	6	E	BE	{0,1}	1

NOTES (on table 6)

Code Notes

1   VDC INTEGER PRECISION: has 1 parameter:

     P1: (integer) VDC integer precision; legal values are 16, 24 or 32; the value 8 is not permitted.

2   VDC REAL PRECISION: has 3 parameters:

     P1: (enumerated) form of representation for real values: valid values are:

         0   floating point format
         1   fixed point format

     P2: (integer) field width for exponent or whole part (including 1 bit for sign)
     P3: (integer) field width for fraction or fractional part

     Legal combinations of values are:

PI	P2	P3	Result
0	9	23	32-bit floating point
0	12	52	64-bit floating point
1	16	16	32-bit fixed point
1	32	32	64-bit fixed point

3   AUXILIARY COLOUR: has 1 parameter; its form depends on COLOUR SELECTION MODE:

     P1: (colour) auxiliary colour. For direct colour selection, a 3-tuple of red, green and blue integer (I) values with range specified by the COLOUR PRECISION element; default is the device-dependent foreground colour. For indexed colour selection, an index (CIX), at COLOUR INDEX PRECISION, into the COLOUR TABLE; default is 0.

4   TRANSPARENCY: has 1 parameter:

     P1: (enumerated) on-off indicator: valid values are:

         0   off: auxiliary colour background is required
         1   on: transparent background is required

5   CLIP RECTANGLE: has 2 parameters:

P1: (point) first point
P2: (point) second point

6   CLIP INDICATOR: has 1 parameter:

P1: (enumerated) clip indicator: valid values are:

0   off
1   on

7.6 Graphical Primitive Elements

TABLE 7. Encoding of Graphical Primitive Elements

Element Class 4	Element Id	Parameter Type	Parameter List Length	Parameter Range	Default
POLYLINE	1	nP	nBP	VDCR	n/a
DISJOINT POLYLINE	2	nP	nBP	VDCR	n/a
POLYMARKER	3	nP	nBP	VDCR	n/a
TEXT	4	P,E,S	BP+BE+ BS	VDCR,{0,1}, SR	n/a
RESTRICTED TEXT	5	2VDC,P, E,S	2VDC,BP+ BE+BS	2VDCR,VDCR, {0,1},SR	n/a
APPEND TEXT	6	E,S	BE+BS	{0,1},SR	n/a
POLYGON	7	nP	nBP	VDCR	n/a
POLYGON SET	8	n(P,E)	n(BP+BE)	VDCR,0..3	n/a
CELL ARRAY	9	3P,3I, E,CLIST	3BP+3BI+ BE+nBCO	VDCR,+IR, {0,1},COR	n/a
GENERALIZED DRAWING PRIMITIVE	10	I,I,nP,D	2BI+nBP+ BS	IR,++IR, VDCR,SR	n/a
RECTANGLE	11	2P	2BP	VDCR	n/a
CIRCLE	12	P,VDC	BP+BVDC	VDCR, ++VDCR	n/a
CIRCULAR ARC 3 POINT	13	3P	3BP	VDCR	n/a
CIRCULAR ARC 3 POINT CLOSE	14	3P,E	3BP+BE	VDCR,{0,1}	n/a
CIRCULAR ARC CENTRE	15	P,4VDC, VDC	BP+4BVDC+ BVDC	VDCR,VDCR, ++VDCR	n/a
CIRCULAR ARC CENTRE CLOSE	16	P,4VDC, VDC,E	BP+4BVDC+ BVDC+BE	VDCR,VDCR, ++VDCR,{0,1}	n/a
ELLIPSE	17	3P	3BP	VDCR	n/a
ELLIPTICAL ARC	18	3P,4VDC	3BP+4BVDC	VDCR,VDCR	n/a
ELLIPTICAL ARC CLOSE	19	3P,4VDC, E	3BP+4BVDC+ BE	VDCR,VDCR, {0,1}	n/a

NOTES (on table 7)

Code Notes

1   POLYLINE: has a variable parameter list:

      P1-Pn: (point) n (X,Y) polyline vertices

2   DISJOINT POLYLINE: has a variable parameter list:

      P1-Pn: (point) n (X,Y) line segment endpoints

3   POLYMARKER: has a variable parameter list:

P1-Pn: (point) n (X,Y) marker positions

4   TEXT: has 3 parameters:

P1: (point) text position
P2: (enumerated) final/not-final flag: valid values are:

    0   not final
    1   final

P3: (string) text string

5   RESTRICTED TEXT: has 5 parameters:

P1: (vdc) delta width
P2: (vdc) delta height
P3: (point) text position
P4: (enumerated) final/not-final flag: valid values are:

    0   not final
    1   final

P5: (string) text string

6   APPEND TEXT: has 2 parameters:

P1: (enumerated) final/not-final flag: valid values are:

    0   not final
    1   final

P2: (string) text string

7   POLYGON: has a variable parameter list:

P1-Pn: (point) n (X,Y) polygon vertices

8   POLYGON SET: has a variable parameter list of pairs of values, each of which has the following form:

P(i):(point) (X,Y) polygon vertex
P(i+1): (enumerated) edge out flag, indicating closures and edge visibility: valid values are

    0   invisible
    1   visible
    2   close, invisible
    3   close, visible

9   CELL ARRAY: has 8 parameters:

P1: (point) corner point P
P2: (point) corner point Q
P3: (point) corner point R
P4: (integer) nx
P5: (integer) ny
P6: (integer) local colour precision: valid values are 0, 1, 2, 4, 8, 16, 24, and 32. If the value is zero (the 'default colour precision indicator' value), the COLOUR (INDEX) PRECISION for the picture indicates the precision with which the colour list is encoded. If the value is non-zero, the precision with which the colour data is encoded is given by the value.
P7: (enumerated) cell representation mode: valid values are:

    0   run length list mode

      1   packed list mode

P8: (colour list) array of cell colour values.

If the COLOUR SELECTION MODE is 'direct', the values will be direct colour values. If the COLOUR SELECTION MODE is 'indexed', the values will be indexes into the COLOUR TABLE.

If the cell representation mode is 'packed list', the colour values are represented by rows of values, each row starting on a word boundary. If the cell representation mode is 'run length', the colour list values are represented by rows broken into runs of constant colour; each row starts on a word boundary. Each list item consists of a cell count (integer) followed by a colour value.

10   GENERALIZED DRAWING PRIMITIVE: has a variable parameter list:

P1: (integer) GDP identifier
P2: (integer) n, number of points in 'list of points'
P3-P(n+2): (point array) list of points
P(n+3)...: (data record) GDP data record

The parameter P2 is required to determine where the coordinate data ends and the data record begins. Data records are bound as strings in this encoding.

11   RECTANGLE: has 2 parameters:

P1: (point) first corner
P2: (point) second corner

12   CIRCLE: has 2 parameters:

P1: (point) centre of circle
P2: (vdc) radius of circle

13   CIRCULAR ARC 3 POINT: has 3 parameters:

P1: (point) starting point
P2: (point) intermediate point
P3: (point) ending point

14   CIRCULAR ARC 3 POINT CLOSE: has 4 parameters:

P1: (point) starting point
P2: (point) intermediate point
P3: (point) ending point
P4: (enumerated) type of arc closure: valid values are:

    0   pie closure
    1   chord closure

15   CIRCULAR ARC CENTRE: has 6 parameters:

P1: (point) centre of circle
P2: (vdc) delta X for start vector
P3: (vdc) delta Y for start vector
P4: (vdc) delta X for end vector
P5: (vdc) delta Y for end vector
P6: (vdc) radius of circle

16   CIRCULAR ARC CENTRE CLOSE: has 7 parameters:

P1: (point) centre of circle
P2: (vdc) delta X for start vector
P3: (vdc) delta Y for start vector
P4: (vdc) delta X for end vector

P5: (vdc) delta Y for end vector
P6: (vdc) radius of circle
P7: (enumerated) type of arc closure: valid values are:

    0   pie closure
    1   chord closure

17   ELLIPSE: has 3 parameters:

P1: (point) centre of ellipse
P2: (point) endpoint of first conjugate diameter
P3: (point) endpoint of second conjugate diameter

18   ELLIPTICAL ARC: has 7 parameters:

P1: (point) centre of ellipse
P2: (point) endpoint for first conjugate diameter
P3: (point) endpoint for second conjugate diameter
P4: (vdc) delta X for start vector
P5: (vdc) delta Y for start vector
P6: (vdc) delta X for end vector
P7: (vdc) delta Y for end vector

19   ELLIPTICAL ARC CLOSE: has 8 parameters:

P1: (point) centre of ellipse
P2: (point) endpoint for first conjugate diameter
P3: (point) endpoint for second conjugate diameter
P4: (vdc) delta X for start vector
P5: (vdc) delta Y for start vector
P6: (vdc) delta X for end vector
P7: (vdc) delta Y for end vector
P8: (enumerated) type of arc closure: valid values are:

    0   pie closure
    1   chord closure

7.7 Attribute Elements

TABLE 8. Encoding of Attribute Elements

Element Class 5	Element Id	Parameter Type	Parameter List Length	Parameter Range	Default
LINE BUNDLE INDEX	1	IX	BIX	+IXR	1
LINE TYPE	2	IX	BIX	IXR	1
LINE WIDTH	3	VDC or R	BVDC or BR	++VDCR or ++RR	see below
LINE COLOUR	4	CO	BCO	COR	see below
MARKER BUNDLE INDEX	5	IX	BIX	+IXR	1
MARKER TYPE	6	IX	BIX	IXR	3
MARKER SIZE	7	VDC or R	BVDC or BR	++VDCR or ++RR	see below
MARKER COLOUR	8	CO	BCO	COR	see below
TEXT BUNDLE INDEX	9	IX	BIX	+IXR	1
TEXT FONT INDEX	10	IX	BIX	+IXR	1
TEXT PRECISION	11	E	BE	0..2	0
CHARACTER EXPANSION FACTOR	12	R	BR	+RR	1.0
CHARACTER SPACING	13	R	BR	RR	0.0
TEXT COLOUR	14	CO	BCO	COR	see below
CHARACTER HEIGHT	15	VDC	BVDC	++VDCR	see below
CHARACTER ORIENTATION	16	4VDC	4BVDC	VDCR	0,1,1,0
TEXT PATH	17	E	BE	0..3	0
TEXT ALIGNMENT	18	2E, R,R	2BE+ 2BR	0..4, 0..6, 2RR	0,0, 0.0,0.0
CHARACTER SET INDEX	19	IX	BIX	+IXR	1
ALTERNATE CHARACTER SET INDEX	20	IX	BIX	+IXR	1

(continued)

TABLE 8. Encoding of Attribute Elements (continued).

FILL BUNDLE INDEX	21	IX	BIX	+IXR	1
INTERIOR STYLE	22	E	BE	ER	0
FILL COLOUR	23	CO	BCO	COR	see below
HATCH INDEX	24	IX	BIX	IXR	1
PATTERN INDEX	25	IX	BIX	+IXR	1
EDGE BUNDLE INDEX	26	IX	BIX	+IXR	1
EDGE TYPE	27	IX	BIX	IXR	1
EDGE WIDTH	28	VDC or R	BVDC or BR	++VDCR or ++RR	see below
EDGE COLOUR	29	CO	BCO	COR	see below
EDGE VISIBILITY	30	E	BE	{0,1}	0
FILL REFERENCE POINT	31	P	BP	VDCR	0,0
PATTERN TABLE	32	IX,3I, nx*nyCO	BIX+3BI+ nx*nyBCO	+IXR,+IR COR	1,(1,1), 0,0
PATTERN SIZE	33	4VDC	4BVDC	VDCR	see below
COLOUR TABLE	34	CI,nCD	BCI+nBCD	CIR,CDR	see below
ASPECT SOURCE FLAGS	35	n(E,E)	n(2BE)	(0..17), {0,1}	(n,0), n=0..17

NOTES (on table 8)

Code Notes

1   LINE BUNDLE INDEX: has 1 parameter:

      P1: (index) line bundle index

2   LINE TYPE: has 1 parameter:

      P1: (index) line type: the following values are standardized:

              1   solid
              2   dash
              3   dot
              4   dash-dot
              5   dash-dot-dot
              negative for private use

3   LINE WIDTH: has 1 parameter:

      P1: (vdc) line width (if size specification is 'absolute') or (real) line width scale factor (if size specification is 'scaled'). Default is 1.0 for 'scaled', or for 'absolute' 0.001 times the longest side of the default VDC EXTENT.

4   LINE COLOUR: has 1 parameter; its form depends on COLOUR SELECTION MODE:

      P1: (colour) line colour. For direct colour selection, a 3-tuple of red, green and blue integer (I) values with range specified by the COLOUR PRECISION element; default is the device-dependent foreground colour. For indexed colour selection, an index (CIX), at COLOUR INDEX PRECISION, into the COLOUR TABLE; default is 1.

5   MARKER BUNDLE INDEX: has 1 parameter:

P1: (index) marker bundle index

6 MARKER TYPE: has 1 parameter:

P1: (index) marker type: the following values are standardized:

 1 dot
 2 plus
 3 asterisk
 4 circle
 5 cross
 negative for private use

7 MARKER SIZE: has 1 parameter:

P1: (vdc) marker size (if size specification is 'absolute') or (real) marker size scale factor (if size specification is 'scaled'). Default is 1.0 for 'scaled', or for 'absolute' 0.01 times the longest side of the default VDC EXTENT.

8 MARKER COLOUR: has 1 parameter; its form depends on COLOUR SELECTION MODE:

P1: (colour) marker colour. For direct colour selection, a 3-tuple of red, green and blue integer (I) values with range specified by the COLOUR PRECISION element; default is the device-dependent foreground colour. For indexed colour selection, an index (CIX), at COLOUR INDEX PRECISION, into the COLOUR TABLE; default is 1.

9 TEXT BUNDLE INDEX: has 1 parameter:

P1: (index) text bundle index

10 TEXT FONT INDEX: has 1 parameter:

P1: (index) text font index

11 TEXT PRECISION: has 1 parameter:

P1: (enumerated) text precision: valid values are:

 0 string
 1 character
 2 stroke

12 CHARACTER EXPANSION FACTOR: has 1 parameter:

P1: (real) character expansion factor

13 CHARACTER SPACING: has 1 parameter:

P1: (real) additional inter-character space

14 TEXT COLOUR: has 1 parameter; its form depends on COLOUR SELECTION MODE:

P1: (colour) text colour. For direct colour selection, a 3-tuple of red, green and blue integer (I) values with range specified by the COLOUR PRECISION element; default is the device-dependent foreground colour. For indexed colour selection, an index (CIX), at COLOUR INDEX PRECISION, into the COLOUR TABLE; default is 1.

15 CHARACTER HEIGHT: has 1 parameter:

P1: (vdc) character height. Default is 1.0 for 'scaled', or for 'absolute' 0.01 times the longest side of the default VDC EXTENT.

16 CHARACTER ORIENTATION: has 4 parameters:

P1: (vdc) X character up component
P2: (vdc) Y character up component
P3: (vdc) X character base component
P4: (vdc) Y character base component

17   TEXT PATH: has 1 parameter:

    P1: (enumerated) text path: valid values are:

      0   right
      1   left
      2   up
      3   down

18   TEXT ALIGNMENT: has 4 parameters:

    P1: (enumerated) horizontal alignment: valid values ares:

      0   normal
      1   left
      2   centre
      3   right
      4   continuous horizontal

    P2: (enumerated) vertical alignment

      0   normal
      1   top
      2   cap
      3   half
      4   base
      5   bottom
      6   continuous vertical

    P3: (real) continuous horizontal alignment
    P4: (real) continuous vertical alignment

19   CHARACTER SET INDEX: has 1 parameter:

    P1: (index) character set index

20   ALTERNATE CHARACTER SET INDEX: has 1 parameter:

    P1: (index) alternate character set index

21   FILL BUNDLE INDEX: has 1 parameter:

    P1: (index) fill bundle index

22   INTERIOR STYLE: has 1 parameter:

    P1: (enumerated) interior style: valid values are:

      0   hollow
      1   solid
      2   pattern
      3   hatch
      4   empty
      negative for private use

23   FILL COLOUR: has 1 parameter; its form depends on COLOUR SELECTION MODE:

    P1: (colour) fill colour. For direct colour selection, a 3-tuple of red, green and blue integer (I) values with range specified by the COLOUR PRECISION element; default is the device-dependent foreground colour. For indexed colour selection, an index (CIX), at COLOUR INDEX PRECISION, into the COLOUR TABLE; default is 1.

24   HATCH INDEX: has 1 parameter

    P1: (index) hatch index: the following values are standardized:

      1   horizontal
      2   vertical

3   positive slope
4   negative slope
5   combined vertical and horizontal slant
6   combined left and right slant.
negative for private use

25   PATTERN INDEX: has 1 parameter

P1: (index) pattern index

26   EDGE BUNDLE INDEX: has 1 parameter:

P1: (index) edge bundle index

27   EDGE TYPE: has 1 parameter:

P1: (integer) edge type: the following values are standardized:

1   solid
2   dash
3   dot
4   dash-dot
5   dash-dot-dot
negative for private use

28   EDGE WIDTH: has 1 parameter:

P1: (vdc) edge width (if size specification is absolute) or (real) edge width scale factor (if size specification is scaled). Default is 1.0 for 'scaled', or for 'absolute' 0.001 times the longest side of the default VDC EXTENT.

29   EDGE COLOUR: has 1 parameter; its form depends on COLOUR SELECTION MODE:

P1: (colour) edge colour. For direct colour selection, a 3-tuple of red, green and blue integer (I) values with range specified by the COLOUR PRECISION element; default is the device-dependent foreground colour. For indexed colour selection, an index (CIX), at COLOUR INDEX PRECISION, into the COLOUR TABLE; default is 1.

30   EDGE VISIBILITY: has 1 parameter:

P1: (enumerated) edge visibility: valid values are:

0   off
1   on

31   FILL REFERENCE POINT: has 1 parameters:

P1: (point) fill reference point

32   PATTERN TABLE: has 5 parameters:

P1: (index) pattern table index
P2: (integer) nx, the dimension of colour array in the direction of the PATTERN SIZE width vector
P3: (integer) ny, the dimension of colour array in the direction of the PATTERN SIZE height vector
P4: (integer) local colour precision: valid values are 0, 1, 2, 4, 8, 16, 24, and 32. If the value is zero (the 'default colour precision indicator' value), the COLOUR (INDEX) PRECISION for the picture indicates the precision with which the colour list is encoded. If the value is non-zero, the precision with which the colour data is encoded is given by the value.
P5: (colour array) pattern definition

33   PATTERN SIZE: has 4 parameters:

P1: (vdc) pattern height vector, x component
P2: (vdc) pattern height vector, y component

P3: (vdc) pattern width vector, x component
P4: (vdc) pattern width vector, y component

The default pattern size is 0,dy,dx,0, where dx and dy are respectively the height and width of the VDC extent.

34   COLOUR TABLE: has 2 parameters:

P1: (colour index) starting colour table index
P2: (direct colour list) list of direct colour values (red,green,blue 3-tuples)

35   ASPECT SOURCE FLAGS: has up to 18 parameter-pairs, corresponding to each attribute that may be bundled; each parameter-pair contains the ASF type and the ASF value:

(enumerated) ASF type; valid values are:

0   line type ASF
1   line width ASF
2   line colour ASF
3   marker type ASF
4   marker size ASF
5   marker colour ASF
6   text font index ASF
7   text precision ASF
8   character expansion factor ASF
9   character spacing ASF
10   text colour ASF
11   interior style ASF
12   fill colour ASF
13   hatch index ASF
14   pattern index ASF
15   edge type ASF
16   edge width ASF
17   edge colour ASF

(enumerated) ASF value; valid values are:

0   individual
1   bundled

7.8 Escape Element

TABLE 9. Encoding of Escape Element

Element Class 6	Element Id	Parameter Type	Parameter List Length	Parameter Range	Default
ESCAPE	1	I,D	BI+BS	IR,SR	n/a

NOTES (on table 9)

Code Notes

1   ESCAPE: has 2 parameters:

P1: (integer) escape identifier
P2: (data record) escape data record; data records are bound as strings in this encoding.

7.9 External Elements

TABLE 10. Encoding of External Elements

Element Class 7	Element Id	Parameter Type	Parameter List Length	Parameter Range	Default
MESSAGE	1	E,S	BE+BS	{0,1},SR	0 or n/a
APPLICATION DATA	2	I,D	BI+BS	IR,SR	n/a

NOTES (on table 10)

Code Notes

1   MESSAGE: has 2 parameters:

  P1: (enumerated) action-required flag: valid values are:

    0   no action
    1   action

  P2: (string) message string

2   APPLICATION DATA: has 2 parameters:

  P1: (integer) identifier
  P2: (data record) application data record; data records are bound as strings in this encoding.

8 Defaults

The following are the defaults for the binary encoding.

REAL PRECISION | Fixed point; whole part 16 bits; fractional part 16 bits.

INTEGER PRECISION | 16 bits

COLOUR PRECISION | 1 octet (per colour component)

COLOUR INDEX PRECISION | 1 octet

INDEX PRECISION | 16 bits

VDC REAL PRECISION | Fixed point; whole part 16 bits; fractional part 16 bits.

VDC INTEGER PRECISION | 16 bits

VDC EXTENT | If VDC TYPE is REAL, default VDC EXTENT is (0.0,0.0) , (0.9999..,0.9999..).

If VDC TYPE is INTEGER, default VDC EXTENT is (0,0) , (32767,32767).

COLOUR VALUE EXTENT: | minimum is (0,0,0) and maximum is (255,255,255).

9 Conformance

A metafile fully conforms to this encoding if it meets the following criteria.

— Each metafile element in this part is coded in the manner described.

— Private (non-standard) metafile elements are all coded using the GENERALIZED DRAWING PRIMITIVE or ESCAPE metafile elements as appropriate. Opcodes reserved for future standardization are not used to code private (non-standard) metafile elements.

— Private (non-standard) values of index parameters are all coded using negative integers.

— Values specified as being "reserved for registration or future standardization" are not used unless their meaning has been registered or standardized.

Inclusion of non-graphical data in the metafile should be accomplished with the APPLICATION DATA element. Communication with external agencies, such as device operators, should be accomplished with the MESSAGE element.

A conforming metafile may include, within the string parameters of TEXT, RESTRICTED TEXT, and APPEND TEXT elements, as well as string parameters within the data records of GENERALIZED DRAWING PRIMITIVE (GDP) elements (their admissibility here depends upon the particular GDP definition), the ISO 2022 controls for designating and invoking G-sets. This is an alternative way, in addition to CHARACTER SET INDEX, by which character sets for displaying text strings may be selected. However, the use of ISO 2022 controls within text strings is implementation dependent. Metafile interpreters are not required to respond correctly to the ISO 2022 controls for designating and invoking G-sets when those controls occur within text strings of a TEXT, RESTRICTED TEXT, APPEND TEXT, or GDP element.

Part 3

ANNEXES

A Formal Grammar

NOTE - This annex is not part of the Standard; it is included for information purposes only.

This annex provides explanation of the terminal symbols specified in annex A of part 1 of this Standard.

Opcodes are encoded as two integers specifying the element class and element identifier. The element classes are listed in table 2 and the element identifiers in tables 3, 4, 5, 6, 7 and 8. The full list of class and element codes is given in annex C. For example:

<METAFILE VERSION>	::=	1 1 <parameter list length>
<METAFILE DESCRIPTION>	::=	1 2 <parameter list length>
<parameter list length>	::=	<integer> {encoded as described in clause 4}

The enumerated types are 16-bit signed integers. The other terminal symbols are described in detail in clause 5. A reference to the relevant tables is given here.

<integer>	::=	two's complement integer {See clause 5}
<real>	::=	<floating point real> ¦ <fixed point real> {See clause 5}
<coordinate>	::=	<integer>(2) ¦ <real>(2)
<VDC value>	::=	<integer> ¦ <real>
<string>	::=	<length> <character>* {See table 1}
<character>	::=	8-bit characters, or multiples of 8 bits, depending on the character set {See 5.3}.
<colour index>	::=	<unsigned integer> {See table 1}
<red green blue>	::=	<unsigned integer>(3) {See table 1}

Character substitution is not used in this binding.

The following operands are described in table 4:

<index precision value>
<integer precision value>
<colour precision value>
<colour index precision value>

The following operands are described in table 6.

<integer VDC precision value>

Formal Grammar

The following operand is described in table 7 (under CELL ARRAY):

<colour list>

The following operand value is described in table 7 (under CELL ARRAY):

<default col precision indicator>

The following operand is described in table 9:

<escape data list>

B Examples

NOTE - This annex is not part of the Standard; it is included for information purposes only.

The following simple examples illustrate the use of the binary encoding of the CGM. All precisions used are the default values.

B.1 Example 1 : BEGIN METAFILE 'Example 1'

	15 14 13 12	11 10 9 8 7 6 5	4 3 2 1 0
Header :	0	1	10
Length :	9		'E'
Name :	'x'		'a'
	'm'		'p'
	'l'		'e'
	' '		'1'

B.2 Example 2 : BEGIN PICTURE 'Test'

	15 14 13 12	11 10 9 8 7 6 5	4 3 2 1 0
Header :	0	3	5
Length :	4		'T'
Name :	'e'		's'
(pad)	't'		0

46   ANSI X3.122 - 1986 Part 3

Example 3 : POLYLINE from 0,2 to 1,3 to 2,1 to 0,2

: POLYLINE from 0,2 to 1,3 to 2,1 to 0,2

```
15 14 13 12 11 10  9  8  7  6  5  4  3  2  1  0
+-----------+------------------+------------------+
|     4     |        1         |       16         |
+-----------+------------------+------------------+
|                     0                           |
+-------------------------------------------------+
|                     2                           |
+-------------------------------------------------+
|                     1                           |
+-------------------------------------------------+
|                     3                           |
+-------------------------------------------------+
|                     2                           |
+-------------------------------------------------+
|                     1                           |
+-------------------------------------------------+
|                     0                           |
+-------------------------------------------------+
|                     2                           |
+-------------------------------------------------+
```

4 : TEXT 'Hydrogen' at 0,1

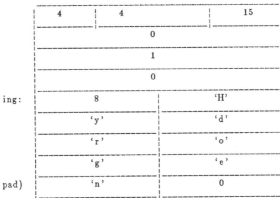

```
   15 14 13 12 11 10  9  8  7  6  5  4  3  2  1  0
  +-----------+------------------+------------------+
  |     4     |        4         |       15         |
  +-----------+------------------+------------------+
  |                     0                           |
  +-------------------------------------------------+
  |                     1                           |
  +-------------------------------------------------+
  |                     0                           |
  +-------------------------------------------------+
ing:  |         8          |        'H'         |
  +--------------------+--------------------+
  |        'y'         |        'd'         |
  +--------------------+--------------------+
  |        'r'         |        'o'         |
  +--------------------+--------------------+
  |        'g'         |        'e'         |
  +--------------------+--------------------+
pad)  |        'n'         |         0          |
  +--------------------+--------------------+
```

Example 5 : Partitioned POLYLINE with 50 points   Examples

B.5 Example 5 : Partitioned POLYLINE with 50 points

```
                 15 14 13 12 11 10  9  8  7  6  5  4  3  2  1  0
                +----------------------------------------------+
Header :        |      4     |     1              |     31     |
                +------------+--------------------+------------+
Long(cont.):    | 1 |              120                         |
                +---+------------------------------------------+
Point(1):       |                  x(1)                        |
                +----------------------------------------------+
                |                  y(1)                        |
                +----------------------------------------------+


Point(30):      |                  x(30)                       |
                +----------------------------------------------+
                |                  y(30)                       |
                +----------------------------------------------+
Long(final):    | 0 |              80                          |
                +---+------------------------------------------+
Point(31):      |                  x(31)                       |
                +----------------------------------------------+
                |                  y(31)                       |
                +----------------------------------------------+


Point(50):      |                  x(50)                       |
                +----------------------------------------------+
                |                  y(50)                       |
                +----------------------------------------------+
```

B.6 Example 6 : METAFILE DEFAULT REPLACEMENT linewidth 0.5

```
             15 14 13 12 11 10  9  8  7  6  5  4  3  2  1  0
            ------------------------------------------------
Header:     |    1    |       11        |         4         |
            ------------------------------------------------
Line width: |    5    |       3         |         2         |
            ------------------------------------------------
Value:      |                       0.5                     |
            ------------------------------------------------
```

B.7 Example 7 : Application Data # 655 with 10K octets (chars) of data

```
               15 14 13 12 11 10  9  8  7  6  5  4  3  2  1  0
              ------------------------------------------------
Header:       |    7    |      2        |         31         |
              ------------------------------------------------
Long(final):  | 0|           10244          |
              --------------------------------
ID:           |                  655                         |
              ------------------------------------------------
Length:       |       255       | 0|   10240  ...            |
              ------------------------------------------------
Data record:  | ... length contd.  |      Octet 1           |
              ------------------------------------------------
              |    Octet 2         |      Octet 3           |
              ------------------------------------------------


(pad)         |    Octet 10240     |         0              |
              ------------------------------------------------
```

C List of Binary Encoding Metafile Element Codes

NOTE - This annex is not part of the Standard; it is included for information purposes only.

The following list, arranged in the order of the elements given in part 1 of this Standard, indicates the class and element codes associated with each metafile element. These are the codes used in the METAFILE ELEMENTS LIST element.

Delimiter Elements: Class 0

Class	Element Code	Element Name
0	1	BEGIN METAFILE
0	2	END METAFILE
0	3	BEGIN PICTURE
0	4	BEGIN PICTURE BODY
0	5	END PICTURE

Metafile Descriptor Elements: Class 1

Class	Element Code	Element Name
1	1	METAFILE VERSION
1	2	METAFILE DESCRIPTION
1	3	VDC TYPE
1	4	INTEGER PRECISION
1	5	REAL PRECISION
1	6	INDEX PRECISION
1	7	COLOUR PRECISION
1	8	COLOUR INDEX PRECISION
1	9	MAXIMUM COLOUR INDEX
1	10	COLOUR VALUE EXTENT
1	11	METAFILE ELEMENT LIST
1	12	METAFILE DEFAULTS REPLACEMENT
1	13	FONT LIST
1	14	CHARACTER SET LIST
1	15	CHARACTER CODING ANNOUNCER

Picture Descriptor Elements: Class 2

Class	Element Code	Element Name
2	1	SCALING MODE
2	2	COLOUR SELECTION MODE
2	3	LINE WIDTH SPECIFICATION MODE
2	4	MARKER SIZE SPECIFICATION MODE
2	5	EDGE WIDTH SPECIFICATION MODE
2	6	VDC EXTENT

List of Binary Encoding Metafile Element Codes

2	7	BACKGROUND COLOUR

Control Elements: Class 3

Class	Element Code	Element Name
3	1	VDC INTEGER PRECISION
3	2	VDC REAL PRECISION
3	3	AUXILIARY COLOUR
3	4	TRANSPARENCY
3	5	CLIP RECTANGLE
3	6	CLIP INDICATOR

Graphical Primitive Elements: Class 4

Class	Element Code	Element Name
4	1	POLYLINE
4	2	DISJOINT POLYLINE
4	3	POLYMARKER
4	4	TEXT
4	5	RESTRICTED TEXT
4	6	APPEND TEXT
4	7	POLYGON
4	8	POLYGON SET
4	9	CELL ARRAY
4	10	GENERALIZED DRAWING PRIMITIVE
4	11	RECTANGLE
4	12	CIRCLE
4	13	CIRCULAR ARC 3 POINT
4	14	CIRCULAR ARC 3 POINT CLOSE
4	15	CIRCULAR ARC CENTRE
4	16	CIRCULAR ARC CENTRE CLOSE
4	17	ELLIPSE
4	18	ELLIPTICAL ARC
4	19	ELLIPTICAL ARC CLOSE

Attribute Elements: Class 5

Class	Element Code	Element Name
5	1	LINE BUNDLE INDEX
5	2	LINE TYPE
5	3	LINE WIDTH
5	4	LINE COLOUR
5	5	MARKER BUNDLE INDEX
5	6	MARKER TYPE
5	7	MARKER SIZE
5	8	MARKER COLOUR
5	9	TEXT BUNDLE INDEX

5	10	TEXT FONT INDEX
5	11	TEXT PRECISION
5	12	CHARACTER EXPANSION FACTOR
5	13	CHARACTER SPACING
5	14	TEXT COLOUR
5	15	CHARACTER HEIGHT
5	16	CHARACTER ORIENTATION
5	17	TEXT PATH
5	18	TEXT ALIGNMENT
5	19	CHARACTER SET INDEX
5	20	ALTERNATE CHARACTER SET INDEX
5	21	FILL BUNDLE INDEX
5	22	INTERIOR STYLE
5	23	FILL COLOUR
5	24	HATCH INDEX
5	25	PATTERN INDEX
5	26	EDGE BUNDLE INDEX
5	27	EDGE TYPE
5	28	EDGE WIDTH
5	29	EDGE COLOUR
5	30	EDGE VISIBILITY
5	31	FILL REFERENCE POINT
5	32	PATTERN TABLE
5	33	PATTERN SIZE
5	34	COLOUR TABLE
5	35	ASPECT SOURCE FLAGS

Escape Element: Class 6

Class	Element Code	Element Name
6	1	ESCAPE

External Elements: Class 7

Class	Element Code	Element Name
7	1	MESSAGE
7	2	APPLICATION DATA

ANSI X3.122 - 1986

Information Processing Systems

Computer Graphics

Metafile for the Storage and Transfer
of Picture Description Information

Part 4

Clear Text Encoding

ANSI X3.122 - 1986
Part 4

CONTENTS

0 Introduction

0.1 Purpose of the Clear Text Encoding

The Clear Text Encoding of the Computer Graphics Metafile (CGM) provides a representation of the Metafile syntax that is easy to type, edit and read. It allows a metafile to be edited with any standard text editor, using the internal character code of the host computer system.

0.2 Primary Objectives

a. HUMAN EDITABLE: The Clear Text Encoding should be able to be hand edited or, if desired, hand constructed.

b. HUMAN-FRIENDLY: The Clear Text Encoding should be easy and natural for people to read and edit. Although what is easiest and most natural is a subjective judgment that varies among users, contributing factors such as ease of recognition, ease of remembering, avoidance of ambiguity, and prevention of mistyping have all been considered.

c. MACHINE-READABLE: The Clear Text Encoding should be able to be parsed by software.

d. Suitable for USE IN A WIDE VARIETY OF EDITORS: The Clear Text Encoding should not have any features that make it difficult to edit in normal text editors.

e. Facilitate INTERCHANGE BETWEEN DIVERSE SYSTEMS: The Clear Text Encoding should be encoded in such a way as to maximize the set of systems which can utilize it. No assumptions should be made as to word size or arithmetic modes used to interpret the metafile.

f. Use STANDARDIZED ABBREVIATIONS as much as possible. Where language encoding of other graphics standards have established standard abbreviations, or where common practice in the data processing and graphics industries has established well known abbreviations, these abbreviations are used. In accordance with the principle of "least astonishment", this approach should minimize the time needed to learn to use this encoding.

0.3 Secondary Objectives

Because other CGM encodings are targeted toward CPU efficiency (CGM Binary Encoding) and information density (CGM Character Encoding), these objectives are considered of secondary importance for the CGM Clear Text Encoding.

0.4 Relationship to Other Standards

The set of characters required to implement the Clear Text Encoding is a subset of those included in national versions of ISO 646. Any character set that can be mapped to and from that subset may be used to implement the encoding.

For certain elements, the CGM defines value ranges as being reserved for registration. The values and their meanings will be defined using the established procedures (see part 1, 4.11.)

0.5 Status of Annexes

The annexes do not form an integral part of the Standard but are included to provide extra information and explanation.

1 Scope and Field of Application

1.1 Scope

This part of this Standard specifies a clear text encoding of the Computer Graphics Metafile. For each of the elements specified in part 1 of this Standard, a clear text encoding is specified. Allowed abbreviations are specified. The overall format of the metafile and the means by which comments may be interspersed in the metafile is specified.

1.2 Field of Application

This encoding of the CGM allows metafiles to be created and maintained in a form which is simple to type, easy to edit and convenient to read.

2 References

ISO 646   7-bit coded character set for information processing interchange.

ISO 2022   Information processing - ISO 7-bit and 8-bit coded character sets - Code extension techniques.

3 Notational Conventions

Unbracketed strings are terminals of this grammar which appear exactly, subject to the notes on case and null characters given below.

Bracketed strings are either non-terminals (with further productions given), character symbol names (such as COMMA), or parameters of the CGM element in the form <x:y> (see part 1 of the Standard for further explanation of these items).

"::=" is read as "becomes" or "is realized as".

<...>*   = star closure (0 or more occurrences).
<...>+   = plus closure (1 or more occurrences).
<...>o   = optional (exactly 0 or 1 occurrences).
<x:y>   = parameter type x with meaning y
<x¦y>   = exactly one of x or y
{...}   = a comment (not part of the production)

SPACES are used for readability in the grammar description; SPACES in the actual metafile are indicated through the separator productions given below.

The metasymbols used in describing the grammar do not appear in the actual metafile.

4 Entering and Leaving the Metafile Environment

4.1 Generic Clear Text and Instantiations

The Clear Text Encoding is described in a generic fashion that permits it to be used with any character set capable of representing those characters enumerated in the Character Repertoire (see part 1, 4.7.6). An instantiation of the Clear Text Encoding is specified by defining the character set and coding technique to be used (for example, standard national character sets based on ISO 646, non-standard character sets such as EBCDIC, etc).

It is recommended that an instantiation of the Clear Text Encoding bound to the standard national character set based on ISO 646 be used in order to maximize portability of Clear Text metafiles between diverse systems. This also provides an encoding which can be incorporated into an ISO 2022 text environment as a complete code, to permit intermixing of text and graphics for applications which place a high priority on human readability.

4.2 Implicitly Entering the Metafile Environment

The Clear Text coding environment may be entered implicitly by agreement between the interchanging parties. This is suitable only if there is not to be any interchange with services using other coding techniques, and if it is known by prior agreement which instantiation of the syntax is being used.

4.3 Designating and Invoking the CGM Coding Environment from ISO 2022

For interchange with services using the code extension techniques of ISO 2022, the (standard national version) ISO 646 instantiation of the CGM Clear Text Encoding may be designated and invoked from the ISO 2022 environment by the following escape sequence:

ESC 2/5 F

where ESC is the bit combination 1/11, and F refers to a bit combination that will be assigned by the ISO Registration Authority for ISO 2375.

The first bit combination occurring after this escape sequence will then represent the beginning of a CGM metafile element or one of the "soft separators" or "null characters" defined below.

The following escape sequence may be used to return to the ISO 2022 coding environment:

ESC 2/5 4/0

This not only returns to the ISO 2022 coding environment, but also restores the designation and invocation of coded character sets to the state that existed prior to entering the ISO 646 CGM coding environment with the ESC 2/5 F sequence. (The terms "designation" and "invocation" are defined in ISO 2022.)

It is permissable to make transitions between ISO 2022 and the metafile environment between pictures in the metafile as well as between metafiles. The state of the metafile interpreter and the state of the ISO 2022 environment are maintained separately and not stacked. The state of the metafile interpreter before BEGIN METAFILE or after END METAFILE is undefined, and sending a picture without a preceding BEGIN METAFILE and metafile descriptor is nonconforming interchange.

5 Metafile Format

A metafile in the Clear Text Encoding consists of a stream of characters forming a series of elements, each of which starts with an element name and ends with one of the element delimiters, either the SLASH character (also known as SLANT or SOLIDUS) or the SEMICOLON character. (Note: these characters do not act as element delimiters when occurring within the bounds of a string parameter, as defined below.)

5.1 Character Repertoire

In order to achieve objective (e) of sub-clause 0.2, the character repertoire of the Clear Text Encoding will be limited to those characters enumerated below, except for string parameters, which may contain any characters from the repertoire described in part 1 of this Standard (see part 1, 4.7.6).

- Upper-case characters:
  "A", "B", "C", "D", "E", "F", "G", "H", "I",
  "J", "K", "L", "M", "N", "O", "P", "Q", "R",
  "S", "T", "U", "V", "W", "X", "Y", "Z"
- Lower-case characters: (see note 1)
  "a", "b", "c", "d", "e", "f", "g", "h", "i",
  "j", "k", "l", "m", "n", "o", "p", "q", "r",
  "s", "t", "u", "v", "w", "x", "y", "z"
- Digits:
  "0", "1", "2", "3", "4", "5", "6", "7", "8", "9"
- " " (SPACE character)
- "+" (PLUS character)
- "-" (MINUS character)
- "#" (NUMBER SIGN)
- ";" (SEMICOLON character)
- "/" (SLASH, SLANT, or SOLIDUS character)
- "(" (LEFT or OPEN PARENTHESIS character)
- ")" (RIGHT or CLOSE PARENTHESIS character)
- "," (COMMA character)
- "." (DECIMAL POINT or PERIOD character)
- "'" (APOSTROPHE or SINGLE QUOTE character)
- """ (DOUBLE QUOTE character)
- "_" (UNDERSCORE character) (see note 2)
- "$" (DOLLAR SIGN or CURRENCY symbol) (see note 2)
- "%" (PERCENT SIGN character)

NOTE 1: lower-case characters are considered to be the same as upper-case characters, when occurring outside of string parameters. Any combination of lower-case and upper-case characters may be used within an element or enumerated parameter name.

NOTE 2: The UNDERSCORE and DOLLAR SIGN symbols are defined as "null characters" within this encoding. They may appear anywhere within the metafile, and are mandated to have no effect on parsing (outside of string parameters). They are available for the generator or editor of the metafile to use in enhancing readability of tokens.

EXAMPLES: The following are all equivalent: linetype, LINETYPE, LineType, line_type, $LINE-TYPE, L_I_N_E$T_Y_P_E. The following are all equivalent: 123456, $123456, 123_456, $123_456, $12$34$56.

Those control characters that are format effectors (BACKSPACE, CARRIAGE RETURN, LINEFEED, NEWLINE, HORIZONTAL TAB, VERTICAL TAB, and FORMFEED) are permitted in the metafile, but are treated as SPACE characters (that is, as soft delimiters) by the metafile interpreter whenever they occur outside of a string parameter. They may be used to assist in formatting the metafile to improve its readability. (The effect of such format effectors within string parameters is as defined in part 1 of this Standard.) A metafile written in the Clear Text Encoding is considered to be non-conforming interchange if it includes characters other than those listed in the repertoire and the format effectors (outside of string parameters). Implementation-dependent extensions which require use of characters other than the above should be embedded in the string parameters of the ESCAPE, MESSAGE, or APPLICATION DATA elements, or in comments.

Note that the code set of the characters is not fixed by this Standard. In order to accomplish the objective of editability, it is permitted to encode the Clear Text Encoding using the character set codes native to the system. It is presumed that standard conversion facilities can be used in translating Clear Text CGM files from one system's character set codes to another, consistent with the treatment of other text files being transferred between systems. It is recommended that the ISO 646 codes be used to encode Clear Text metafiles for transport between diverse systems.

Null characters or format effectors outside of text strings which do not exist in the target system's encoding may be dropped in such translation, and lower-case letters translated to upper case as necessary, without altering the information content of the metafile. Likewise, the two statement delimiter characters are interchangeable and may be changed in such a translation without affecting the information content of the metafile. The two string delimiter characters are interchangeable, but any translation shall correctly handle the possible occurrence of either string delimiter character within the string parameter.

5.2 Separators

5.2.1 Element Separators

<TERM> ::= <OPTSEP> <SLASH ¦ SEMICOLON> <OPTSEP>

The SEMICOLON and SLASH characters may be used interchangeably to delimit elements in a Clear Text metafile. These elements do not, however, terminate an element when they occur within a string parameter, as described below.

The elements of the metafile are not terminated by the ends of records, as indicated by control characters such as CR (carriage return) or LF (linefeed). Multiple elements may exist on one line, and any element may extend over multiple lines.

5.2.2 Parameter Separators

The following productions are used in the Clear Text Encoding for parameter separators:

<SEPCHAR>	::=	<SPACE ¦ CARRIAGE RETURN ¦ LINEFEED ¦ HORIZONTAL TAB ¦ VERTICAL TAB ¦ FORMFEED>
<SOFTSEP>	::=	<SEPCHAR>+
<OPTSEP>	::=	<SEPCHAR>*
<HARDSEP>	::=	<OPTSEP> <COMMA> <OPTSEP>
<SEP>	::=	<SOFTSEP> ¦ <HARDSEP>

Most commands require a SOFTSEP after the element name (e.g., at least one space). This permits element names to be formed from a mixture of alpha and numeric characters.

The separator between parameters is usually a SEP. This format permits omission of parameters. (Two consecutive COMMAs indicate an omitted parameter.)

Since the enclosing APOSTROPHE or DOUBLE QUOTE character sufficiently delineates string parameters, and the statement delimiter SLASH also sets off the data on either side of it, the separators between these characters and adjacent parameters or element names are optional (OPTSEP).

SEPCHAR characters are not permitted within a name (element or enumerated type), or within the representation of a numeric parameter. Any place where a SEPCHAR is permitted (other than inside a string parameter), an arbitrary number of SEPCHARs may be used.

5.2.3 Comments in the Metafile

Comments may be included in a Clear Text metafile, to enhance its readability and usefulness. Some uses of comments might be to document hand-edited changes to the metafile, or as "notes to one's self" made while reading a metafile. To include other forms of nongraphical information in the metafile, it is suggested that the APPLICATION DATA element be used. If it is desired to convert a Clear Text metafile to one of the other encodings, comments may be either dropped or converted to APPLICATION DATA elements.

Comments are encoded as a series of printing characters and <SEPCHAR>s surrounded by "%" (PER-CENT SIGN) characters. The text of the comment may not include this comment delimiter character.

Comments may be included any place that a separator may be used, and are equivalent to a <SOFT-SEP>; they may be replaced by a SPACE character in parsing, without affecting the meaning of the metafile.

5.3 Encoding of Parameter Types

5.3.1 Integer-Bound Types

INTEGERS, INTEGER COORDINATES, INDICES, and the components of COLOUR DIRECT parameters are all bound to signed integers, indicated in the encoding as I.

<I>	::=	<decimal integer> ¦ <based integer>
<decimal integer>	::=	<sign>o <digit>+
<sign>	::=	<PLUS SIGN> ¦ <MINUS SIGN>
<based integer>	::=	<sign>o <base> <NUMBER SIGN> <extended digit>+
<base>	::=	2 ¦ 3 ¦ 4 ¦ 5 ¦ 6 ¦ 7 ¦ 8 ¦ 9 ¦ 10 ¦ 11 ¦ 12 ¦ 13 ¦ 14 ¦ 15 ¦ 16
<digit>	::=	0 ¦ 1 ¦ 2 ¦ 3 ¦ 4 ¦ 5 ¦ 6 ¦ 7 ¦ 8 ¦ 9
<extended digit>	::=	<digit> ¦ A ¦ B ¦ C ¦ D ¦ E ¦ F ¦ a ¦ b ¦ c ¦ d ¦ e ¦ f

The null characters are permitted within numbers, but are not shown in the productions for simplicity.

A decimal integer has an optional sign and at least one digit. If the sign appears, it immediately precedes the number with no intervening SPACE (or other <SEPCHAR>) characters allowed.

A based integer has an optional sign, a base (an unsigned integer in the range 2..16 inclusive, represented in base 10), a "#", and a string of one or more extended digits. If the sign appears, it immediately precedes the number with no intervening SPACE (or other <SEPCHAR>) characters allowed. The extended digits used shall be valid for the base named or the metafile is not conforming; e.g., for base 8 the digits "8", "9", etc. are not valid, for base 2 only the digits "0" and "1" are valid, and so forth. Case is not significant for the extended digits.

If the sign is omitted for either form, the number is considered non-negative.

Both the base and the <extended digit>+ are interpreted as unsigned numbers, and the final result negated if a MINUS SIGN preceded the number. No assumptions should be made as to the word size of the metafile interpreter, or whether the underlying arithmetic is one's complement, two's complement, or sign-magnitude. For example, -1 would be encoded in hexadecimal as -16#1, -16#0001, etc. rather than 16#FFFF. Of course, metafiles may be created utilizing prior knowledge of the intended target machine, but any such assumptions will limit the portability of the metafile and are discouraged.

Examples:

> 0, 007, -5, +123_456
> The following are equivalent: 65535, 16#FFFF, 16#ffff, 8#177777, 2#1111_1111_1111_1111
> The following are equivalent: -32_768, -16#8000, -8#100000, -2#10000000_00000000

Interpretation of numerically bound parameters will be "free field", that is, there is an implied radix point to the right of the rightmost digit, and neither leading nor trailing spaces are significant. Leading zeroes are not significant.

5.3.2 Real-Bound Types

REALS and REAL COORDINATES are bound to real numbers, indicated in the encoding as R. These are written as either explicit-point numbers or scaled-real numbers (or decimal integers, where appropriate).

<R>	::=	< explicit-point number > ¦
		< scaled-real number> ¦
		< decimal integer >
< explicit-point number >	::=	<sign>o
		<
		<<digit>+ <PERIOD> <digit>*>
		¦
		<<digit>* <PERIOD> <digit>+>
		>
< scaled-real number>	::=	<body> < E ¦ e > <exponent>
<body>	::=	<explicit-point number> ¦
		<decimal integer>
<exponent>	::=	<decimal integer>

The interpretation of the scaled-real number is the same as standard scientific notation (similar to FORTRAN "E" format), where the number represented by <body> is multiplied by 10 taken to the power <exponent>.

There shall be at least one digit in an explicit-point number and in the body of a scaled-real number, which in the case of a single-digit number may appear on either side of the radix point. It is recommended but not required that there be at least one digit before the radix point, for numbers with only a fractional part. Zero may be encoded as "0·", ".0", "0.0", "0", etc., although the second form is not recommended.

In the case of a scaled-real number (one where an "E" or "e" appears), at least one digit shall appear in the <exponent>. No SPACE or other <SEPCHAR> characters are permitted between the <body> and the "E" or "e", or between the "E" or "e" and the <exponent>.

The interpretation of parameters bound to this data type will be "free field", that is, if there is an explicit radix point, it sets the radix point of the internal representation, and neither leading nor trailing spaces or zeroes are significant. If the radix point is omitted, it is implied to be at the right of the rightmost digit of the explicit-point number or of the <body> of the scaled-real number. Thus, decimal I-format numbers are permitted to appear in a conforming metafile for parameters bound to real numbers when there is no fractional part.

For real numbers in all formats, the only permitted base of representation is base 10.

If the <sign> ("+" or "-") is omitted, the number is assumed to be non-negative. If the sign is present, it immediately precedes the body of the number, with no SPACE (or other <SEPCHAR>) characters allowed between it and the leftmost digit or radix point of the body of the number.

COMMA, SPACE and other <SEPCHAR> characters are not permitted within a number, but <NULLCHAR> characters are permitted (and have no effect on parsing).

Examples:
        3.14159
        7.853982E-7
        271828e-5
        42
        -.04321 (not recommended form)
        -0.043_21
        42E2

5.3.3 String-bound Types

STRING parameters are represented as character strings immediately surrounded by a matched pair of either APOSTROPHE (SINGLE QUOTE) or DOUBLE QUOTE characters.

If an APOSTROPHE is required in a string delimited with APOSTROPHES, it is represented by two adjacent APOSTROPHES at that position in the string. Likewise, if a DOUBLE QUOTE character is required in a string delimited with DOUBLE QUOTE characters, it is represented by two adjacent DOUBLE QUOTE characters. For example, the following are equivalent

    TEXT 0 0 FINAL "Murphy's Law: ""If it can go wrong, it will.""";
    TEXT 0 0 FINAL 'Murphy''s Law: "If it can go wrong, it will."';

DATA RECORD data type is represented as a string in this encoding.

STRING parameters are indicated in the encoding as S.

5.3.4 Enumerated Types

Enumerated types are bound to names, similarly to the element names. Where an implementation wishes to support private enumerated type values, these shall be encoded as the letters "PRIV" followed by a string of <alpha | digit | null character>*.

5.3.5 Derived Types

In addition to the I, R, and S parameter formats, the following abbreviations are used as shorthand for the productions shown.

VDC ::= <I> | <R>   { coordinate data, depending on VDC TYPE }

<RED GREEN BLUE> ::= <I:RED> <SEP> <I:GREEN> <SEP> <I:BLUE>

K ::= <I> | <RED GREEN BLUE> { colour value, depending on COLOUR SELECTION MODE }

POINTREC ::= <VDC> <SEP> <VDC>

P   ::=   <POINTREC> |
           < <LEFT PAREN> <OPTSEP> <POINTREC> <OPTSEP> <RIGHT PAREN> >

{ COORDINATE in VDC space. Parentheses are optional. If they are used, they shall group exactly two VDC numbers. The parenthesized form is intended to aid readability of the metafile. If there are not two numbers in each parenthesized group, the metafile is non-conforming interchange. Any attempted error recovery or exception handling which a metafile interpreter may use in this situation is neither defined nor constrained by this Standard. A metafile interpreter need not use the parentheses in parsing; in this case, they are treated as SPACE characters rather than as NULL characters (i.e., they act as soft delimiters). }

V ::= <VDC> | <R>
{ Used for line width, marker size, and edge width, this parameter type depends on the corresponding SPECIFICATION MODE element in the picture descriptor. }

5.4 Forming Names

The approach was taken of selecting abbreviations for words used to name elements and enumeration types in the CGM, and concatenating them in order.

5.4.1 Words Deleted

ANNOUNCER	FACTOR	SPECIFICATION
AREA	INDICATOR	
BUNDLE	REPLACEMENT	
CIRCULAR	SELECTION	

5.4.2 Words Added

INCREMENTAL

5.4.3 Words Used Unabbreviated

ABSTRACT	END	PLUS
ACTION	ESCAPE	POLYGON
ARC	FILL	PRIVATE
ARRAY	FINAL	REAL
BASE	FONT	RIGHT
BASIC	HALF	SCALED
BIT	HATCH	SET
BODY	HEIGHT	SIZE
BOTTOM	HOLLOW	SOLID
BUNDLED	INDEX	START
CAP	INDEXED	STRING
CELL	INTEGER	STROKE
CHORD	LEFT	STYLE
CIRCLE	LINE	TABLE
CLIP	LIST	TEXT
CLOSE	MARKER	TRANSPARENCY
CODING	MESSAGE	TYPE
DATA	METRIC	UP
DEFAULTS	MODE	VALUE
DIRECT	NO	VDC

DOWN	NOT	VERSION
DRAWING	OFF	WIDTH
EDGE	ON	3 (numeral three)
ELLIPSE	PATH	
EMPTY	PIE	

5.4.4 Abbreviations

ABSOLUTE	ABS
ALIGNMENT	ALIGN
ALTERNATE	ALT
APPEND	APND
APPLICATION	APPL
ASPECT SOURCE FLAG(S)	ASF
AUXILIARY	AUX
BACKGROUND	BACK
BEGIN	BEG
CENTRE	CTR
CHARACTER	CHAR
COLOUR	COLR
CONTINUOUS	CONT
CONTROL	CTRL
DESCRIPTION	DESC
DISJOINT	DISJT
EIGHT	8
ELEMENT	ELEM
ELLIPTICAL	ELLIP
EXPANSION	EXPAN
EXTENDED	EXTD
GENERALIZED DRAWING PRIMITIVE	GDP
INCREMENTAL	INCR
INDIVIDUAL	INDIV
INTERIOR	INT
INVISIBLE	INVIS
MAXIMUM	MAX
METAFILE	MF
NORMAL	NORM
ORIENTATION	ORI
PATTERN	PAT
PICTURE	PIC
POINT	PT
POLYLINE	LINE
POLYMARKER	MARKER
PRECISION	PREC
RECTANGLE	RECT
REFERENCE	REF
RESTRICTED	RESTR
SCALING	SCALE
SEVEN	7
SPACING	SPACE
VISIBILITY	VIS
VISIBLE	VIS

5.4.5 The Derived Element Names

Metafile Element	Element Name	Notes
BEGIN METAFILE	BEGMF	
END METAFILE	ENDMF	
BEGIN PICTURE	BEGPIC	
BEGIN PICTURE BODY	BEGPICBODY	
END PICTURE	ENDPIC	
METAFILE VERSION	MFVERSION	
METAFILE DESCRIPTION	MFDESC	
VDC TYPE	VDCTYPE	
INTEGER PRECISION	INTEGERPREC	
REAL PRECISION	REALPREC	
INDEX PRECISION	INDEXPREC	
COLOUR PRECISION	COLRPREC	
COLOUR INDEX PRECISION	COLRINDEXPREC	
MAXIMUM COLOUR INDEX	MAXCOLRINDEX	
COLOUR VALUE EXTENT	COLRVALUEEXT	
METAFILE ELEMENT LIST	MFELEMLIST	
METAFILE DEFAULTS REPLACEMENT	BEGMFDEFAULTS	(1)
	ENDMFDEFAULTS	
FONT LIST	FONTLIST	
CHARACTER SET LIST	CHARSETLIST	
CHARACTER CODING ANNOUNCER	CHARCODING	
SCALING MODE	SCALEMODE	
COLOUR SELECTION MODE	COLRMODE	
LINE WIDTH SPECIFICATION MODE	LINEWIDTHMODE	
MARKER SIZE SPECIFICATION MODE	MARKERSIZEMODE	
EDGE WIDTH SPECIFICATION MODE	EDGEWIDTHMODE	
VDC EXTENT	VDCEXT	
BACKGROUND COLOUR	BACKCOLR	
VDC INTEGER PRECISION	VDCINTEGERPREC	
VDC REAL PRECISION	VDCREALPREC	
AUXILIARY COLOUR	AUXCOLR	
TRANSPARENCY	TRANSPARENCY	
CLIP RECTANGLE	CLIPRECT	
CLIP INDICATOR	CLIP	
POLYLINE	LINE	(2)
	INCRLINE	
DISJOINT POLYLINE	DISJTLINE	(2)
	INCRDISJTLINE	
POLYMARKER	MARKER	(2)
	INCRMARKER	
TEXT	TEXT	
RESTRICTED TEXT	RESTRTEXT	
APPEND TEXT	APNDTEXT	
POLYGON	POLYGON	(2)
	INCRPOLYGON	
POLYGON SET	POLYGONSET	(2)
	INCRPOLYGONSET	
CELL ARRAY	CELLARRAY	
GENERALIZED DRAWING PRIMITIVE	GDP	
RECTANGLE	RECT	
CIRCLE	CIRCLE	
CIRCULAR ARC 3 POINT	ARC3PT	
CIRCULAR ARC 3 POINT CLOSE	ARC3PTCLOSE	

CIRCULAR ARC CENTRE	ARCCTR
CIRCULAR ARC CENTRE CLOSE	ARCCTRCLOSE
ELLIPSE	ELLIPSE
ELLIPTICAL ARC	ELLIPARC
ELLIPTICAL ARC CLOSE	ELLIPARCCLOSE
LINE BUNDLE INDEX	LINEINDEX
LINE TYPE	LINETYPE
LINE WIDTH	LINEWIDTH
LINE COLOUR	LINECOLR
MARKER BUNDLE INDEX	MARKERINDEX
MARKER TYPE	MARKERTYPE
MARKER SIZE	MARKERSIZE
MARKER COLOUR	MARKERCOLR
TEXT BUNDLE INDEX	TEXTINDEX
TEXT FONT INDEX	TEXTFONTINDEX
TEXT PRECISION	TEXTPREC
CHARACTER EXPANSION FACTOR	CHAREXPAN
CHARACTER SPACING	CHARSPACE
TEXT COLOUR	TEXTCOLR
CHARACTER HEIGHT	CHARHEIGHT
CHARACTER ORIENTATION	CHARORI
TEXT PATH	TEXTPATH
TEXT ALIGNMENT	TEXTALIGN
CHARACTER SET INDEX	CHARSETINDEX
ALTERNATE CHARACTER SET INDEX	ALTCHARSETINDEX
FILL BUNDLE INDEX	FILLINDEX
INTERIOR STYLE	INTSTYLE
FILL COLOUR	FILLCOLR
HATCH INDEX	HATCHINDEX
PATTERN INDEX	PATINDEX
EDGE BUNDLE INDEX	EDGEINDEX
EDGE TYPE	EDGETYPE
EDGE WIDTH	EDGEWIDTH
EDGE COLOUR	EDGECOLR
EDGE VISIBILITY	EDGEVIS
FILL REFERENCE POINT	FILLREFPT
PATTERN TABLE	PATTABLE
PATTERN SIZE	PATSIZE
COLOUR TABLE	COLRTABLE
ASPECT SOURCE FLAGS	ASF
ESCAPE	ESCAPE
MESSAGE	MESSAGE
APPLICATION DATA	APPLDATA

NOTES -

1) This element is implemented by a pair of Clear Text element names in the metafile, one to begin defaults replacement and the second to end defaults replacement.

2) These elements have point list parameters, the points of which may be represented either with absolute or incremental coordinates. For each of these elements there are two possible Clear Text element names, one corresponding to absolute coordinate representation and one to incremental coordinate representation — the element name used shall correspond to the coordinate representation of point list.

6 Encoding the CGM Elements

6.1 Allowable Productions in Clear Text Metafiles

All productions given below which do not appear in the table in 5.4.5 are merely "syntactic shorthand" for describing element productions, and may not appear by themselves in the metafile. For example, CELLROW is a handy way to describe a piece of the CELL ARRAY element, and is not a primitive in itself.

6.2 Encoding Delimiter Elements

BEGIN METAFILE	::= BEGMF
	    <OPTSEP>
	    <S:NAME>
	    <TERM>

END METAFILE	::= ENDMF <TERM>

BEGIN PICTURE	::= BEGPIC
	    <OPTSEP>
	    <S:PICTURENAME>
	    <TERM>

BEGIN PICTURE BODY	::= BEGPICBODY <TERM>
END PICTURE	::= ENDPIC <TERM>

6.3 Encoding Metafile Descriptor Elements

METAFILE VERSION	::= MFVERSION
	    <SOFTSEP>
	    <I:VERSION>
	    <TERM>

METAFILE DESCRIPTION	::= MFDESC
	    <OPTSEP>
	    <S:DESCRIPTION>
	    <TERM>

VDC TYPE	::= VDCTYPE
	    <SOFTSEP>
	    < INTEGER ¦ REAL >
	    <TERM>

INTEGER PRECISION	::= INTEGERPREC
	    <SOFTSEP>
	    <I:MININT>
	    <SEP>
	    <I:MAXINT>
	    <TERM>

Discussion: the most negative and most positive integers (base 10) are given. These parameters are interpreted independently of all precisions currently set.

REAL PRECISION   ::= REALPREC
                              <SOFTSEP>
                              <F:MINREAL>
                              <SEP>
                              <F:MAXREAL>
                              <SEP>
                              <I:DIGITS>
                              <TERM>

Discussion: The parameters of this element are interpreted independently of all precisions currently set. The MINREAL and MAXREAL are signed real numbers giving the representable range of numbers. The DIGITS parameter gives the minimum number of DECIMAL DIGITS of accuracy assumed, and is of key importance in preventing roundoff error when the incremental forms of output primitives are used. (Note: this choice of format was patterned after the floating_point_constraint of the Ada programming language.)

INDEX PRECISION   ::= INDEXPREC
                              <SOFTSEP>
                              <I:MININT>
                              <SEP>
                              <I:MAXINT>
                              <TERM>
(see INTEGERPREC for discussion)

COLOUR PRECISION   ::= COLRPREC
                              <SOFTSEP>
                              <I:MAXCOMPONENT>
                              <TERM>

Discussion: Colour direct values are 3*I. COLOUR PRECISION gives a single integer range, 0..MAXCOMPONENT, within which each of the three compontents is contained. The parameters are interpreted independently of any currently set precisions.

COLOUR INDEX PRECISION   ::= COLRINDEXPREC
                              <SOFTSEP>
                              <I:MAXINT>
                              <TERM>

Discussion: the smallest colour index is 0. The most positive integer that may occur in a colour index parameter is given. This parameter is interpreted independently of all precisions currently set. See MAXCOLRINDEX.

MAXIMUM COLOUR INDEX   ::= MAXCOLRINDEX
                              <SOFTSEP>
                              <I:MAXINDEXVALUE>
                              <TERM>

Discussion: this element gives the colour table size (minus 1) assumed by the metafile writer. It shall be less than or equal to the <MAXINT> parameter of COLRINDEXPREC.

COLOUR VALUE EXTENT   ::= COLRVALUEEXT
                              <SOFTSEP>
                              <BLACK:RED GREEN BLUE>
                              <SEP>

<WHITE:RED GREEN BLUE>
<TERM>

Discussion: the 3-tuples corresponding to black and white are given.

METAFILE ELEMENT LIST   ::= MFELEMLIST
                             <OPTSEP>
                             <S:ELEMENTNAMES>
                             <TERM>

Discussion: The string parameter consists of a list of element names separated by <SEP>. In addition, the words DRAWINGPLUS and DRAWINGSET may be used in this string, as defined in part 1 of the Standard.

METAFILE DEFAULTS REPLACEMENT ::= BEGMFDEFAULTS
                                  <TERM>

                                  <ELEMENT>+

                                  ENDMFDEFAULTS
                                  <TERM>

Discussion: Between the two bracketing elements, applicable CGM elements appear using the same format as described elsewhere in this section.

FONT LIST   ::= FONTLIST
                                 <OPTSEP>
                                 <S:FONTNAME>
                                 < <SEP> <S:FONTNAME> >*
                                 <TERM>

CHARACTER SET LIST   ::= CHARSETLIST
                                 <CHARSETDESIGNATOR>+
                                 <TERM>

CHARSETDESIGNATOR   ::= <SOFTSEP>
                                 < STD94 |
                                   STD96 |
                                   STD94MULTIBYTE |
                                   STD96MULTIBYTE |
                                   COMPLETECODE >
                                 < <OPTSEP> | <SEP> >
                                 <S:TAIL>

CHARACTER CODING ANNOUNCER   ::= CHARCODING
                                  < BASIC7BIT |
                                    BASIC8BIT |
                                    EXTD7BIT |
                                    EXTD8BIT >
                                  <TERM>

6.4 Encoding Picture Descriptor Elements

```
SCALING MODE                    ::=  SCALEMODE
                                     <SOFTSEP>
                                     < ABSTRACT ¦ METRIC >
                                     <SEP>
                                     <R:SCALEFACTOR>
                                     <TERM>

COLOUR SELECTION MODE  ::=  COLRMODE
                                     <SOFTSEP>
                                     < INDEXED ¦ DIRECT >
                                     <TERM>

LINE WIDTH SPECIFICATION MODE     ::=  LINEWIDTHMODE
                                     <SOFTSEP>
                                     < ABSTRACT ¦ SCALED >
                                     <TERM>

MARKER SIZE SPECIFICATION MODE  ::=  MARKERSIZEMODE
                                     <SOFTSEP>
                                     < ABSTRACT ¦ SCALED >
                                     <TERM>

EDGE WIDTH SPECIFICATION MODE     ::=  EDGEWIDTHMODE
                                     <SOFTSEP>
                                     < ABSTRACT ¦ SCALED >
                                     <TERM>

VDC EXTENT                      ::=  VDCEXT
                                     <SOFTSEP>
                                     <P:FIRSTCORNER>
                                     <SEP>
                                     <P:SECONDCORNER>
                                     <TERM>

BACKGROUND COLOUR       ::=  BACKCOLR
                                     <SOFTSEP>
                                     <RED GREEN BLUE>
                                     <TERM>
```

6.5 Encoding Control Elements

```
VDC INTEGER PREC        ::=  VDCINTEGERPREC
                                     <SOFTSEP>
                                     <I:MININT>
                                     <SEP>
                                     <I:MAXINT>
                                     <TERM>
     (see INTEGERPREC for discussion)

VDC REAL PRECISION      ::=  VDCREALPREC
                                     <SOFTSEP>
                                     <F:MINREAL>
```

```
                              <SEP>
                              <F:MAXREAL>
                              <SEP>
                              <I:DIGITS>
                              <TERM>
    (see REALPREC for discussion )
```

```
AUXILIARY COLOUR          ::=  AUXCOLR
                               <SOFTSEP>
                               <K:AUXCOLR>
                               <TERM>
```

```
TRANSPARENCY              ::=  TRANSPARENCY
                               <SOFTSEP>
                               < OFF | ON >
                               <TERM>
```

```
CLIP RECTANGLE            ::=  CLIPRECT
                               <SOFTSEP>
                               <P:FIRSTCORNER>
                               <SEP>
                               <P:SECONDCORNER>
                               <TERM>
```

```
CLIP INDICATOR            ::=  CLIP
                               <SOFTSEP>
                               < OFF | ON >
                                            <TERM>
```

6.6 Encoding Graphical Primitive Elements

```
POLYLINE                  ::=  < LINE
                               <POINTLIST>
                               <TERM>
                               >
                               |
                               < INCRLINE
                               <INCRPOINTLIST>
                               <TERM>
                               >
```

Discussion: LINE and INCRLINE always have at least two points.

```
POINTLIST                 ::=  <SOFTSEP>
                               <P:POINT>
                               < <SEP> <P:POINT> >+
```

```
INCRPOINTLIST             ::=  <SOFTSEP>
                               <P:FIRSTPOINT>
                               <DELTA>+
```

```
DELTA                     ::=  < <SEP> <DELTAPAIR> > |
                               < <SEP> <LEFT PAREN> <DELTAPAIR> <RIGHT PAREN> >
```

```
DELTAPAIR                 ::=  <OPTSEP>
                               <VDC:DELTAX>
```

                                \<SEP\>
                                \<VDC:DELTAY\>
                                \<OPTSEP\>

Discussion: absolute coordinates correspond in a straightforward way to the functionality definition, but incremental coordinates can realize a significant savings in this encoding. Even more important, incremental coordinates are high on the priority list for user-friendly graphics, which is a high priority objective of this encoding.

```
DISJOINT POLYLINE        ::=  < DISJTLINE
                                <SOFTSEP>
                                <P:POINT>
                                <SEP>
                                <P:POINT> >
                                < <SEP>
                                  <P:POINT>
                                  <SEP>
                                  <P:POINT> >*
                                <TERM>
                              >
                              |
                              < INCRDISJTLINE
                                <SOFTSEP>
                                <P:POINT>
                                <DELTA>
                                < <DELTA> <DELTA> >*
                                <TERM>
                              >
```

Discussion: DISJTLINE and INCRDISTJTLINE shall have pairs of points, and at least one pair.

```
POLYMARKER               ::=  < MARKER
                                <POINTLIST>
                                <TERM>
                              >
                              |
                              < INCRMARKER
                                <INCRPOINTLIST>
                                <TERM>
                              >

TEXT                     ::=  TEXT
                                <SOFTSEP>
                                <P:TEXTLOCATION>
                                <SEP>
                                <TEXTPIECE>

RESTRICTED TEXT          ::=  RESTRTEXT
                                <SOFTSEP>
                                <VDC:MAXWIDTH>
                                <SEP>
                                <VDC:MAXHEIGHT>
                                <SEP>
                                <P:TEXTLOCATION>
                                <SEP>
```

```
                            <TEXTPIECE>

APPEND TEXT        ::= APNDTEXT
                            <SOFTSEP>
                            <TEXTPIECE>

TEXTPIECE          ::= < NOTFINAL ¦ FINAL >
                            < <OPTSEP> ¦ <HARDSEP> >
                            <S:TEXTSTRING>
                            <TERM>

POLYGON            ::= < POLYGON
                            <SOFTSEP>
                            <P:POINT>
                            <SEP>
                            <P:POINT>
                            < <SEP> <P:POINT> >+
                            <TERM>
                          >
                          ¦
                          < INCRPOLYGON
                            <SOFTSEP>
                            <P:POINT>
                            <DELTA>
                            <DELTA>+
                            <TERM>
                          >

POLYGON SET        ::= < POLYGONSET
                            <SOFTSEP>
                            <FLAGGEDPOINT>
                            <SEP>
                            <FLAGGEDPOINT>
                            < <SEP> <FLAGGEDPOINT> >+
                            <TERM>
                          >
                          ¦
                          < INCRPOLYGONSET
                            <SOFTSEP>
                            <FLAGGEDPOINT>
                            < <FLAGGEDDELTA> <SEP> >+
                            <FLAGGEDDELTA>
                            <TERM>
                          >
```

Discussion: for all POLYGON-type primitives, at least three points shall be present.

```
EDGEFLAG           ::= INVIS ¦ VIS ¦ CLOSEINVIS ¦ CLOSEVIS

FLAGGEDPOINT       ::= <P:VERTEX> <SEP> <EDGEFLAG>

FLAGGEDDELTA       ::= <DELTA> <SEP> <EDGEFLAG>

CELL ARRAY         ::= CELLARRAY
                            <SOFTSEP>
                            <P:P_POINT>
```

```
                            <SEP>
                            <P:Q_POINT>
                            <SEP>
                            <P:R_POINT>
                            <SEP>
                            <I:NX>
                            <SEP>
                            <I:NY>
                            <LOCLCOLRPREC>
                            <CELLROW>*
                        <TERM>
```

LOCLCOLRPREC　　　　　　 ::= < <SEP> <I:MAXINT> > ¦
　　　　　　　　　　　　　　　< <SEP> <I:MAXCOMPONENT> > ¦
　　　　　　　　　　　　　　　< <SEP> <I:0> >

Discussion: the LOCLCOLRPREC takes the form of COLRINDEXPREC or COLRPREC, depending on whether the COLRMODE is <INDEXED> or <DIRECT>, respectively. The value 0 is the 'default precision indicator, and denotes that the precision currently in effect is to be used.

CELLROW　　　　　　　　　 ::= < <SEP> <CELLLIST> > ¦
　　　　　　　　　　　　　　　< <SEP> <LEFT PAREN>
　　　　　　　　　　　　　　　<CELLLIST> <RIGHT PAREN> >

CELLLIST　　　　　　　　　 ::= <NULL> ¦
　　　　　　　　　　　　　　　< <K:CELL> < <SEP> <K:CELL> >* >

Discussion: The parenthesized form of the list is optional. If parentheses are used, they delimit a row of cells. Each row is considered to start from the side defined by the points (P,Q,R) as defined in part 1 of this standard; if a row is not complete, it is left as a metafile-interpreter dependency whether the rest of the row is left untouched or whether the last cell in the row is replicated to fill the row. If there are too many cells in a row, the surplus cells are thrown away. The parentheses may alternately be treated as SPACE characters, in which case the cells are processed sequentially and no row adjusting is done.

The local colour precision parameter, LOCLCOLRPREC, takes the form of the parameter of a colour-precision-setting element, either of COLRPREC or of COLRINDEXPREC, depending on the COLRMODE. Legal values are either legal values of one of those colour precision elements, or all zeros. If the values are all zeros, then the metafile's COLRPREC or COLRINDEXPREC is to be used within the CELLARRAY element as well.

GENERALIZED DRAWING PRIMITIVE　::= GDP
　　　　　　　　　　　　　　　　　　<SOFTSEP>
　　　　　　　　　　　　　　　　　　<I:GDP_IDENTIFIER>
　　　　　　　　　　　　　　　　　　<SEP>
　　　　　　　　　　　　　　　　　　<POINTLIST>
　　　　　　　　　　　　　　　　　　< <OPTSEP> ¦ <HARDSEP> >
　　　　　　　　　　　　　　　　　　<S:DATARECORD>
　　　　　　　　　　　　　　　　<TERM>

RECTANGLE　　　　　　　　 ::= RECT
　　　　　　　　　　　　　　　<SOFTSEP>
　　　　　　　　　　　　　　　<P:FIRSTCORNER>
　　　　　　　　　　　　　　　<SEP>

```
                              <P:SECONDCORNER>
                              <TERM>

CIRCLE                    ::=  CIRCLE
                              <SOFTSEP>
                              <P:CENTRE>
                              <SEP>
                              <VDC:RADIUS>  {non-negative}
                              <TERM>

CIRCULAR ARC 3 POINT      ::=  ARC3PT
                              <3PTARCSPEC>
                              <TERM>

CIRCULAR ARC 3 POINT CLOSE  ::=  ARC3PTCLOSE
                                  <3PTARCSPEC>
                                  <CLOSESPEC>
                                  <TERM>

CLOSESPEC                 ::=  <SEP>
                              < PIE | CHORD >

3PTARCSPEC                ::=  <SOFTSEP>
                              <P:STARTPOINT>
                              <SEP>
                              <P:INTERMEDIATEPOINT>
                              <SEP>
                              <P:ENDPOINT>

CIRCULAR ARC CENTRE       ::=  ARCCTR
                                  <CTRARCSPEC>
                                  <TERM>

CIRCULAR ARC CENTRE CLOSE  ::=  ARCCTRCLOSE
                                  <CTRARCSPEC>
                                  <CLOSESPEC>
                                  <TERM>

CTRARCSPEC                ::=  <SOFTSEP>
                              <P:CENTREPOINT>
                              <ARCBOUNDS>
                              <SEP>
                              <VDC:RADIUS>  {non-negative}

ARCBOUNDS                 ::=  <SEP>
                              <P:STARTVECTOR>
                              <SEP>
                              <P:ENDVECTOR>
```

Discussion: the start and end vectors are given as point records rather than 2*VDC to permit the parenthesized form to be used to represent the vectors.

```
ELLIPSE                   ::=  ELLIPSE
                                  <ELLIPSESPEC>
                                  <TERM>
```

ELLIPSESPEC   ::= <SOFTSEP>
                          <P:CENTRE>
                          <SEP>
                          <P:ENDPOINT_FIRST_CONJUGATE_DIAMETER>
                          <SEP>
                          <P:ENDPOINT_SECOND_CONJUGATE_DIAMETER>

ELLIPTICAL ARC   ::= ELLIPARC
                          <ELLIPSESPEC>
                          <ARCBOUNDS>
                          <TERM>

ELLIPTICAL ARC CLOSE ::= ELLIPARCCLOSE
                          <ELLIPSESPEC>
                          <ARCBOUNDS>
                          <CLOSESPEC>
                          <TERM>

6.7 Encoding Attribute Elements

LINE BUNDLE INDEX   ::= LINEINDEX
                          <SOFTSEP>
                          <I:BUNDLEINDEX> {positive}
                          <TERM>

LINE TYPE   ::= LINETYPE
                          <SOFTSEP>
                          <I:LINETYPE>
                          { 1=solid, 2=dash, 3=dot,
                            4=dash-dot, 5=dash-dot-dot,
                            <0 implementation dependent}
                          <TERM>

LINE WIDTH   ::= LINEWIDTH
                          <SOFTSEP>
                          <V:LINEWIDTH> {non-negative}
                          <TERM>

LINE COLOUR   ::= LINECOLR
                          <SOFTSEP>
                          <K:LINECOLR>
                          <TERM>

MARKER BUNDLE INDEX   ::= MARKERINDEX
                          <SOFTSEP>
                          <I:BUNDLEINDEX> {positive}
                          <TERM>

MARKER TYPE   ::= MARKERTYPE
                          <SOFTSEP>
                          <I:MARKERTYPE>
                          { 1=dot, 2=plus, 3=asterisk, 4=circle
                            5=cross (x), <0 implementation dependent }
                          <TERM>

MARKER SIZE   ::= MARKERSIZE
                                 <SOFTSEP>
                                 <V:MARKERSIZE> {non-negative}
                                 <TERM>

MARKER COLOUR   ::= MARKERCOLR
                                 <SOFTSEP>
                                 <K:MARKERCOLR>
                                 <TERM>

TEXT BUNDLE INDEX   ::= TEXTINDEX
                                 <SOFTSEP>
                                 <I:BUNDLEINDEX> {positive}
                                 <TERM>

TEXT FONT INDEX   ::= TEXTFONTINDEX
                                 <SOFTSEP>
                                 <I:FONTINDEX>   {positive}
                                 <TERM>

TEXT PRECISION   ::= TEXTPREC
                                 <SOFTSEP>
                                 < STRING | CHAR | STROKE >
                                 <TERM>

CHARACTER EXPANSION FACTOR ::= CHAREXPAN
                                 <SOFTSEP>
                                 <R:FACTOR>
                                 <TERM>

CHARACTER SPACING   ::= CHARSPACE
                                 <SOFTSEP>
                                 <R:SPACING>
                                 <TERM>

TEXT COLOUR   ::= TEXTCOLR
                                 <SOFTSEP>
                                 <K:TEXTCOLR>
                                 <TERM>

CHARACTER HEIGHT   ::= CHARHEIGHT
                                 <SOFTSEP>
                                 <VDC:CHARHEIGHT> {non-negative}
                                 <TERM>

CHARACTER ORIENTATION   ::= CHARORI
                                 <SOFTSEP>
                                 <DELTAPAIR> {up vector}
                                 <SEP>
                                 <DELTAPAIR> {base vector}
                                 <TERM>

TEXT PATH   ::= TEXTPATH
                                 <SOFTSEP>
                                 < RIGHT | LEFT | UP | DOWN >
                                 <TERM>

```
TEXT ALIGNMENT          ::= TEXTALIGN
                            <SOFTSEP>
                            < NORMHORIZ | LEFT | CTR | RIGHT | CONTHORIZ >
                            <SEP>
                            < NORMVERT | TOP | CAP | HALF | BASE | BOTTOM |
                               CONTVERT >
                            <SEP>
                            <R:CONTINUOUS_HORIZONTAL>
                            <SEP>
                            <R:CONTINUOUS_VERTICAL)
                            <TERM>

CHARACTER SET INDEX     ::= CHARSETINDEX
                            <SOFTSEP>
                            <I:CHARSETINDEX> {positive}
                            <TERM>

ALTERNATE CHARACTER SET INDEX ::= ALTCHARSETINDEX
                            <SOFTSEP>
                            <I:ALTCHARSETINDEX>  {positive}
                            <TERM>

FILL BUNDLE INDEX       ::= FILLINDEX
                            <SOFTSEP>
                            <I:BUNDLEINDEX> {positive}
                            <TERM>

INTERIOR STYLE          ::= INTSTYLE
                            <SOFTSEP>
                            < HOLLOW | SOLID | PAT | HATCH | EMPTY >
                            <TERM>

FILL COLOUR             ::= FILLCOLR
                            <SOFTSEP>
                            <K:FILLCOLR>
                            <TERM>

HATCH INDEX             ::= HATCHINDEX
                            <SOFTSEP>
                            <I:HATCHINDEX>
                            { 1=horizontal,
                              2=vertical
                              3=positive slope
                              4=negative slope
                              5=horiz/vertical cross
                              6= +/- slope cross
                              <0 implementation dependent}
                            <TERM>

PATTERN INDEX           ::= PATINDEX
                            <SOFTSEP>
                            <I:PATINDEX>  {positive}
                            <TERM>

EDGE BUNDLE INDEX       ::= EDGEINDEX
                            <SOFTSEP>
```

```
                          <I:BUNDLEINDEX> {positive}
                          <TERM>

EDGE TYPE            ::= EDGETYPE
                          <SOFTSEP>
                          <I:EDGETYPE>
                          { 1=solid, 2=dash, 3=dot,
                             4=dash-dot, 5=dash-dot-dot,
                             <0 implementation dependent }
                          <TERM>

EDGE WIDTH          ::= EDGEWIDTH
                          <SOFTSEP>
                          <V:EDGEWIDTH> {non-negative}
                          <TERM>

EDGE COLOUR         ::= EDGECOLR
                          <SOFTSEP>
                          <K:EDGECOLR>
                          <TERM>

EDGE VISIBILITY     ::= EDGEVIS
                          <SOFTSEP>
                          < OFF | ON >
                          <TERM>

FILL REFERENCE POINT ::= FILLREFPT
                          <SOFTSEP>
                          <P:FILLREFPT>
                          <TERM>

PATTERN TABLE       ::= PATTABLE
                          <SOFTSEP>
                          <I:PATINDEX>
                          <SEP>
                          <I:NX>
                          <SEP>
                          <I:NY>
                          <LOCLCOLRPREC>
                          <CELLROW>*
                          <TERM>
```

See CELL ARRAY for discussion of the LOCLCOLRPREC and CELLROW productions.

```
PATTERN SIZE        ::= PATSIZE
                          <SOFTSEP>
                          <DELTAPAIR> {height vector}
                          <SEP>
                          <DELTAPAIR> {width vector}
                          <TERM>

COLOUR TABLE        ::= COLRTABLE
                          <SOFTSEP>
                          <I:STARTINGINDEX>  {non-negative}
                          < <SEP> <RED GREEN BLUE> >*
                          <TERM>
```

ASPECT SOURCE FLAGS   ::=   ASF
                      <SOFTSEP>
                      <ASFPAIR>
                      < <SEP> <ASFPAIR> >*
                      <TERM>

ASFPAIR   ::=   < ASFTYPE >
                      <SEP>
                      < INDIV ¦ BUNDLED >
                      <TERM>

ASFTYPE   ::=   < ASFNAME ¦ PSEUDOASF >

ASFNAME   ::=   < LINETYPE ¦ LINEWIDTH ¦ LINECOLR ¦
                      MARKERTYPE ¦ MARKERSIZE ¦ MARKERCOLR ¦
                      TEXTFONTINDEX ¦ TEXTPREC ¦ CHAREXP ¦
                      CHARSPACE ¦ TEXTCOLR ¦ INTSTYLE ¦
                      FILLCOLR ¦ HATCHINDEX ¦ PATINDEX ¦
                      EDGETYPE ¦ EDGEWIDTH ¦ EDGECOLR >

PSEUDOASF   ::=   < ALL ¦ ALLLINE ¦ ALLMARKER ¦ ALLTEXT ¦
                      ALLFILL ¦ ALLEDGE >

Discussion: The ASF type may either be a valid ASF name, or a pseudo-ASF. If the former, then the name matches the name of the corresponding attribute element. The pseudo-ASFs are a shorthand convenience for setting a number of ASFs at once, and have the meanings:

ALL: set all ASFs as indicated.
ALLLINE: set LINETYPE, LINEWIDTH, and LINECOLR ASFs as indicated.
ALLMARKER: set MARKERTYPE, MARKERSIZE, and MARKERCOLR ASFs as indicated.
ALLTEXT: set TEXTFONTINDEX, TEXTPREC, CHAREXP, CHARSPACE, and TEXTCOLR ASFs as indicated.
ALLFILL: set INTSTYLE, FILLCOLR, HATCHINDEX, and PATINDEX ASFs as indicated.
ALLEDGE: set EDGETYPE, EDGEWIDTH, and EDGECOLR as indicated.

6.8 Encoding Escape Elements

ESCAPE   ::=   ESCAPE
                      <SOFTSEP>
                      <I:IDENTIFIER>
                      < <OPTSEP> ¦ <HARDSEP> >
                      <S:DATARECORD>
                      <TERM>

6.9 Encoding External Elements

MESSAGE   ::=   MESSAGE
                      <SOFTSEP>
                      < NOACTION ¦ ACTION >
                      < <OPTSEP> ¦ <HARDSEP> >
                      <S:MESSAGE_TEXT>

<TERM>

APPLICATION DATA   ::= APPLDATA
                                  <SOFTSEP>
                                  <I:IDENTIFIER>
                                  < <OPTSEP> | <HARDSEP> >
                                  <S:DATARECORD>
                                  <TERM>

7 Clear Text Encoding Defaults

CGM precisions for the Clear Text Encoding:

Non-VDC Reals:

    MINREAL = -32767
    MAXREAL = 32767
    DIGITS   = 4

VDC Precision for Reals:

    MINREAL = 0.0
    MAXREAL = 1.0
    DIGITS   = 4

Non-VDC Integers, Integer VDC:

    MININT   = -32767
    MAXINT   = 32767

Index:

    MININT   = 0
    MAXINT   = 127

Colour Index:

    MAXINT   = 127

Colour Precision:

    MAXCOMPONENT = 255

Colour Value Extent:

    BLACK   = 0,0,0
    WHITE   = 255,255,255

8 Clear Text Encoding Conformance

A metafile conforms to this CGM Clear Text Encoding if it meets the following criteria:

— Each metafile element described in this section is encoded in the manner described.

— Private (non-standard) metafile elements are all encoded using the ESCAPE and GDP metafile elements. New element names that are not enumerated in this Standard are considered nonconforming interchange.

— Private (non-standard) values of index parameters are all encoded using negative integers. When encoding index parameters, a metafile shall not use nonnegative integers to represent private values of index parameters.

— Private (non-standard) values of enumerated parameters are all encoded using the prefix "PRIV" followed by a string of letters and/or digits.

— All characters in the metafile are from the enumerated character repertoire (see 5.1), except for those within a string-type parameter, and format effectors as described in 5.1.

— Numbers are formatted as defined in sections 5.3.1 and 5.3.2. Other forms of numbers are considered nonconforming interchange.

A conforming metafile may include, within the string parameters of TEXT, RESTRICTED TEXT, and APPEND TEXT elements, as well as string parameters within the data records of GENERALIZED DRAWING PRIMITIVE (GDP) elements (their admissibility here depends upon the particular GDP definition), the ISO 2022 controls for designating and invoking G-sets and C-sets, if and only if the appropriate value of CHARACTER CODING ANNOUNCER is included in the Metafile Descriptor. Private coding systems shall likewise be indicated using CHARACTER CODING ANNOUNCER. If this metafile element is not included in the Metafile Descriptor, the string parameters of TEXT, RESTRICTED TEXT, APPEND TEXT, and strings withing GDP data records, shall not include any control codes except format effectors (which are defined to have no standardized effect) for the metafile to be conforming interchange.

Part 4

ANNEXES

A Clear Text Encoding-dependent Formal Grammar

NOTE - This annex is not part of the Standard; it is included for information purposes only.

```
ALPHA   ::=  "A"  |  "B"  |  "C"  |  "D"  |  "E"  |  "F"  |  "G"  |  "H"  |  "I"  |
             "J"  |  "K"  |  "L"  |  "M"  |  "N"  |  "O"  |  "P"  |  "Q"  |  "R"  |
             "S"  |  "T"  |  "U"  |  "V"  |  "W"  |  "X"  |  "Y"  |  "Z"  |
             "a"  |  "b"  |  "c"  |  "d"  |  "e"  |  "f"  |  "g"  |  "h"  |  "i"  |
             "j"  |  "k"  |  "l"  |  "m"  |  "n"  |  "o"  |  "p"  |  "q"  |  "r"  |
             "s"  |  "t"  |  "u"  |  "v"  |  "w"  |  "x"  |  "y"  |  "z"  |

DIGIT   ::=  "0"  |  "1"  |  "2"  |  "3"  |  "4"  |  "5"  |  "6"  |  "7"  |  "8"  |
             "9"
```

NULLCHAR ::= "_" | "$"

CGM opcodes are encoded as element names, and enumerated types are bound to names, as follows:

```
NAME    ::=   <NULLCHAR>*
              <ALPHA>
              < <ALPHA> | <DIGIT> | <NULLCHAR> >*
```

Productions which are encoding-dependent are described in section 3.

B Clear Text Encoding Example

NOTE - This annex is not part of the Standard; it is included for information purposes only.

```
BEGMF 'metafile example';
    mfversion 1; mfdesc'24 January 1984'; vdctype real;
    indexprec -127,127;
    maxcolrindex 7; mfelemlist drawingplus;
    font_list 'Helvetica',
            'Perpetua Bold',
            'CGM_GENERIC: light italic'
            ;
    BEGDEFAULTS;
        VDCEXT 0,0,1,1;
        text_font_index 2;
        int_style solid;
    ENDDEFAULTS;

    % simple picture %
    BEGPIC 'PN 007' ;
        marker_size_mode abs;
    BEGPICBODY;
        % frame %
        line (0,0) (1,0) (1,1) (0,1) (0,0);

        % big dot %
        asf intstyle indiv; circle .5 .5 .3125;


        asf marker_size indiv, marker_type indiv;
        marker_size .005;
        marker_type -3;  % implementation-dependent %
        marker .01 .01
            .5 .5     % note 1 element on several lines %
            .99 .99/
        char_height .04 ;
        text_align ctr, bottom, 0, 0;
        text (.5,0) notfinal "PN 007 is a";
        text_font_index 3 ; apnd_text notfinal ' "silly" ';
        text_font_index 1 ; apnd_text final 'example';
    ENDPIC;
ENDMF ;
```

X3.115-1984 Unformatted 80 Megabyte Trident Pack for Use at 370 tpi and 6000 bpi (General, Physical, and Magnetic Characteristics)

X3.116-1986 Recorded Magnetic Tape Cartridge, 4-Track, Serial 0.250 Inch (6.30 mm) 6400 bpi (252 bpmm), Inverted Modified Frequency Modulation Encoded

X3.117-1984 Printable/Image Areas for Text and Facsimile Communication Equipment

X3.118-1984 Financial Services — Personal Identification Number — PIN Pad

X3.119-1984 Contact Start/Stop Storage Disk, 158361 Flux Transitions per Track, 8.268 Inch (210 mm) Outer Diameter and 3.937 inch (100 mm) Inner Diameter

X3.120-1984 Contact Start/Stop Storage Disk

X3.121-1984 Two-Sided, Unformatted, 8-Inch (200-mm), 48-tpi, Double-Density, Flexible Disk Cartridge for 13 262 ftpr Two-Headed Application

X3.122-1986 Computer Graphics Metafile for the Storage and Transfer of Picture Description Information

X3.124-1985 Graphical Kernel System (GKS) Functional Description

X3.124.1-1985 Graphical Kernel System (GKS) FORTRAN Binding

X3.125-1985 Two-Sided, Double-Density, Unformatted 5.25-inch (130-mm), 48-tpi (1,9-tpmm), Flexible Disk Cartridge for 7958 bpr Use

X3.126-1986 One- or Two-Sided Double-Density Unformatted 5.25-inch (130-mm), 96 Tracks per Inch, Flexible Disk Cartridge

X3.127-1987 Unrecorded Magnetic Tape Cartridge for Information Interchange

X3.128-1986 Contact Start-Stop Storage Disk — 83 000 Flux Transitions per Track, 130-mm (5.118-in) Outer Diameter and 40-mm (1.575-in) Inner Diameter

X3.129-1986 Intelligent Peripheral Interface, Physical Level

X3.130-1986 Intelligent Peripheral Interface, Logical Device Specific Command Sets for Magnetic Disk Drive

X3.131-1986 Small Computer Systems Interface

X3.132-1987 Intelligent Peripheral Interface — Logical Device Generic Command Set for Optical and Magnetic Disks

X3.133-1986 Database Language —NDL

X3.135-1986 Database Language — SQL

X3.136-1986 Serial Recorded Magnetic Tape Cartridge for Information Interchange, Four and Nine Track

X3.139-1987 Fiber Distributed Data Interface (FDDI) Token R Media Access Control (MAC)

X3.140-1986 Open Systems Interconnection — Connection Oriented Transport Layer Protocol Specification

X3.141-1987 Data Communication Systems and Services — Measurement Methods for User-Oriented Performance Evaluation

X3.146-1987 Device Level Interface for Streaming Cartridge and Cassette Tape Drives

X3.147-1987 Intelligent Peripheral Interface — Logical Device Generic Command Set for Magnetic Tapes

X3.153-1987 Open Systems Interconnection — Basic Connectio Oriented Session Protocol Specification

X11.1-1977 Programming Language MUMPS

IEEE 416-1978 Abbreviated Test Language for All Systems (ATLAS)

IEEE 716-1982 Standard C/ATLAS Language

IEEE 717-1982 Standard C/ATLAS Syntax

IEEE 770X3.97-1983 Programming Language PASCAL

IEEE 771-1980 Guide to the Use of ATLAS

ISO 8211-1986 Specifications for a Data Descriptive File for Information Interchange

MIL-STD-1815A-1983 Reference Manual for the Ada Program Language

NBS-ICST 1-1986 Fingerprint Identification — Data Format fo Information Interchange

X3/TRI-82 Dictionary for Information Processing Systems (Technical Report)

American National Standards for Information Processing

X3.1-1976 Synchronous Signaling Rates for Data Transmission

X3.2-1970 Print Specifications for Magnetic Ink Character Recognition

X3.4-1986 Coded Character Sets — 7-Bit ASCII

X3.5-1970 Flowchart Symbols and Their Usage

X3.6-1965 Perforated Tape Code

X3.9-1978 Programming Language FORTRAN

X3.11-1969 General Purpose Paper Cards

X3.14-1983 Recorded Magnetic Tape (200 CPI, NRZI)

X3.15-1976 Bit Sequencing of the American National Standard Code for Information Interchange in Serial-by-Bit Data Transmission

X3.16-1976 Character Structure and Character Parity Sense for Serial-by-Bit Data Communication in the American National Standard Code for Information Interchange

X3.17-1981 Character Set for Optical Character Recognition (OCR-A)

X3.18-1974 One-Inch Perforated Paper Tape

X3.19-1974 Eleven-Sixteenths-Inch Perforated Paper Tape

X3.20-1967 Take-Up Reels for One-Inch Perforated Tape

X3.21-1967 Rectangular Holes in Twelve-Row Punched Cards

X3.22-1983 Recorded Magnetic Tape (800 CPI, NRZI)

X3.23-1985 Programming Language COBOL

X3.25-1976 Character Structure and Character Parity Sense for Parallel-by-Bit Data Communication in the American National Standard Code for Information Interchange

X3.26-1980 Hollerith Punched Card Code

X3.27-1978 Magnetic Tape Labels and File Structure

X3.28-1976 Procedures for the Use of the Communication Control Characters of American National Standard Code for Information Interchange in Specified Data Communication Links

X3.29-1971 Specifications for Properties of Unpunched Oiled Paper Perforator Tape

X3.30-1986 Representation for Calendar Date and Ordinal Date

X3.31-1973 Structure for the Identification of the Counties of the United States

X3.32-1973 Graphic Representation of the Control Characters of American National Standard Code for Information Interchange

X3.34-1972 Interchange Rolls of Perforated Tape

X3.37-1980 Programming Language APT

X3.38-1972 Identification of States of the United States (Including the District of Columbia)

X3.39-1986 Recorded Magnetic Tape (1600 CPI, PE)

X3.40-1983 Unrecorded Magnetic Tape (9-Track 800 CPI, NRZI; 1600 CPI, PE; and 6250 CPI, GCR)

X3.41-1974 Code Extension Techniques for Use with the 7-Bit Coded Character Set of American National Standard Code for Information Interchange

X3.42-1975 Representation of Numeric Values in Character Strings

X3.43-1986 Representations of Local Time of Day

X3.44-1974 Determination of the Performance of Data Communication Systems

X3.45-1982 Character Set for Handprinting

X3.46-1974 Unrecorded Magnetic Six-Disk Pack (General, Physical, and Magnetic Characteristics)

X3.47-1977 Structure for the Identification of Named Populated Places and Related Entities of the States of the United States for Information Interchange

X3.48-1986 Magnetic Tape Cassettes (3.81-mm [0.150-Inch] Tape at 32 bpmm [800 bpi], PE)

X3.49-1975 Character Set for Optical Character Recognition (OCR-B)

X3.50-1986 Representations for U.S. Customary, SI, and Other Units to Be Used in Systems with Limited Character Sets

X3.51-1986 Representations of Universal Time, Local Time Differentials, and United States Time Zone References

X3.52-1976 Unrecorded Single-Disk Cartridge (Front Loading, 2200 BPI) (General, Physical, and Magnetic Requirements)

X3.53-1976 Programming Language PL/I

X3.54-1986 Recorded Magnetic Tape (6250 CPI, Group Coded Recording)

X3.55-1982 Unrecorded Magnetic Tape Cartridge, 0.250 Inch (6.30 mm), 1600 bpi (63 bpmm), Phase encoded

X3.56-1986 Recorded Magnetic Tape Cartridge, 4 Track, 0.250 Inch (6.30 mm), 1600 bpi (63 bpmm), Phase Encoded

X3.57-1977 Structure for Formatting Message Headings Using the American National Standard Code for Information Interchange for Data Communication Systems Control

X3.58-1977 Unrecorded Eleven-Disk Pack (General, Physical, an Magnetic Requirements)

X3.60-1978 Programming Language Minimal BASIC

X3.61-1986 Representation of Geographic Point Locations

X3.62-1987 Paper Used in Optical Character Recognition (OCR) Systems

X3.63-1981 Unrecorded Twelve-Disk Pack (100 Megabytes) (Ge eral, Physical, and Magnetic Requirements)

X3.64-1979 Additional Controls for Use with American Nationa Standard Code for Information Interchange

X3.66-1979 Advanced Data Communication Control Procedures (ADCCP)

X3.72-1981 Parallel Recorded Magnetic Tape Cartridge, 4 Track 0.250 Inch (6.30 mm), 1600 bpi (63 bpmm), Phase Encoded

X3.73-1980 Single-Sided Unformatted Flexible Disk Cartridge (for 6631-BPR Use)

X3.74-1981 Programming Language PL/I, General-Purpose Subs

X3.76-1981 Unformatted Single-Disk Cartridge (Top Loading, 200 tpi 4400 bpi) (General, Physical, and Magnetic Requirement:

X3.77-1980 Representation of Pocket Select Characters

X3.78-1981 Representation of Vertical Carriage Positioning Cha acters in Information Interchange

X3.79-1981 Determination of Performance of Data Communications Systems That Use Bit-Oriented Communication Procedures

X3.80-1981 Interfaces between Flexible Disk Cartridge Drives and Their Host Controllers

X3.82-1980 One-Sided Single-Density Unformatted 5.25-Inch Flexible Disk Cartridge (for 3979-BPR Use)

X3.83-1980 ANSI Sponsorship Procedures for ISO Registration According to ISO 2375

X3.84-1981 Unformatted Twelve-Disk Pack (200 Megabytes)(Ge eral, Physical, and Magnetic Requirements

X3.85-1981 1/2-Inch Magnetic Tape Interchange Using a Self Loading Cartridge

X3.86-1980 Optical Character Recognition (OCR) Inks

X3.88-1981 Computer Program Abstracts

X3.89-1981 Unrecorded Single-Disk, Double-Density Cartridge (Front Loading, 2200 bpi, 200 tpi) (General, Physical, and Magnetic Requirements)

X3.91M-1987 Storage Module Interfaces

X3.92-1981 Data Encryption Algorithm

X3.93M-1981 OCR Character Positioning

X3.94-1985 Programming Language PANCM

X3.95-1982 Microprocessors — Hexadecimal Input/Output, Usin 5-Bit and 7-Bit Teleprinters

X3.96-1983 Continuous Business Forms (Single-Part)

X3.98-1983 Text Information Interchange in Page Image Forma (PIF)

X3.99-1983 Print Quality Guideline for Optical Character Recog tion (OCR)

X3.100-1983 Interface Between Data Terminal Equipment and Data Circuit-Terminating Equipment for Packet Mode Operation with Packet Switched Data Communications Network

X3.101-1984 Interfaces Between Rigid Disk Drive(s) and Host(s)

X3.102-1983 Data Communication Systems and Services — User Oriented Performance Parameters

X3.103-1983 Unrecorded Magnetic Tape Minicassette for Infor tion Interchange, Coplanar 3.81 mm (0.150 in)

X3.104-1983 Recorded Magnetic Tape Minicassette for Informa tion Interchange, Coplanar 3.81 mm (0.150 in), Phase Encoded

X3.105-1983 Data Link Encryption

X3.106-1983 Modes of Operation for the Data Encryption Algor

X3.110-1983 Videotex/Teletext Presentation Level Protocol Sy

X3.111-1986 Optical Character Recognition (OCR) Matrix Char ter Sets for OCR-M

X3.112-1984 14-in (356-mm) Diameter Low-Surface-Friction Magnetic Storage Disk

X3.113-1987 Programming Language FULL BASIC

X3.114-1984 Alphanumeric Machines; Coded Character Sets for Keyboard Arrangements in ANSI X4.23-1982 and X4.22-1983

(Continued on rev

CPSIA information can be obtained
at www.ICGtesting.com
Printed in the USA
BVHW071449051218
534846BV00011B/318/P